Quo Vadis

Quo Vadis

New Directions in the Search for Answers

Robert J. Williams

Writer's Showcase presented by *Writer's Digest*
San Jose New York Lincoln Shanghai

Quo Vadis

New Directions in the Search for Answers

Published by Writer's Showcase presented by *Writer's Digest*
an imprint of iUniverse.com, Inc.

For information address:
iUniverse.com, Inc.
620 North 48th Street
Suite 201
Lincoln, NE 68504-3467
www.iuniverse.com

Front cover photograph appeared in Scientific American's *Understanding Space* © 1984-96 and was taken by Australian astrophotographer David F. Malin.

Photograph © 1993 by Anglo-Australian Observatory.

ISBN: 0-595-09330-2

Printed in the United States of America

Epigraph

Quo vadis is Latin for a question that has been asked of individuals, groups and nations at crucial times throughout history. Now the same question is addressed to all of us. *Quo vadis, Humanity? Where are you going now?*

Contents

Prologue

> I dare to ask that [someday] a child will arise to say:
> On the Meaningless I choose to press a meaning and in the wastes of the Unknowable I choose to be known.
>
> Thornton Wilder, *The Ides of March*

Why isn't the universe teeming with life?

We look out at the cosmos and think that surely we can't be alone. In a universe filled with a hundred billion trillion stars, how could our sun be the only one that has given birth to life? Even if the odds for life were virtually nil—say, 1 in 1,000,000,000,000,000,000—there would still be a hundred thousand solar systems with forms of life ranging from primitive to far more advanced than our species.

So why aren't we receiving any messages?

Part of the answer may be natural catastrophes, such as radical changes in planetary environment or collisions with asteroids, which could have wiped out some of the civilizations that must have emerged over the aeons on other worlds.

But a more universal answer to cosmic silence may be that intelligent life has never found a viable meaning and purpose in its existence.

Life is an experiment, a survival test for the universe's most complex and fragile configurations of matter and energy. When evolution produces a form of intelligent life, the very nature of self-aware, thinking beings drives them to seek an answer to who and what they are and why they exist.

In early stages, the gods answer their questions and give them a leading role in a cosmic plan. But as the creatures climb higher in the tree of knowledge, they find more and more evidence that they exist alone in a vast, uncaring universe. Some retreat into technology's Babylon, where boundless luxury and pleasure soothe the guilt and fears of failure, while those left outside the locked gates assemble their armies and prepare to fight for the city's wealth.

Cosmic silence tells us that other forms of intelligent life have never prevailed over this universe's poison of meaninglessness and never endured for long at the technological level where life acquires the power to destroy itself.

Will life endure on this planet?

We confronted the question for more than four decades in the Cold War, and it certainly was not a grand vision of human destiny that inspired us to step back, at least temporarily, from the nuclear abyss. Now, in our fragmented society, we still have little commitment to the preservation of life. We focus on our immediate needs and interests, such as protecting jobs and business investments, and leave it to our children and grandchildren to worry about any further threats to long-term survival. Let me remind you of just a few of the many ways the story of life on this planet could end:

Global warming...nature's past history of drastic climatic changes in short periods of time...irreversible deterioration of the earth's biosphere...catastrophic upheavals in the earth's thin, unstable crust...a collision with one of the countless asteroids or comets that share the same space with this planet...natural causes or medical research that may release deadly new viruses or bacteria capable of wiping out most or all of the human population...the unleashing of chemical, biological or nuclear weapons by terrorists or nations caught up in regional rivalries.

No doubt you have seen those and other scenarios played out in books and movies. But the science-fiction fantasies will become the inevitable reality, according to scientists, environmentalists and historians, if we don't start making the preservation of life more important than any national, cultural or business interests.

"Not only will mankind not prevail, who says it should even endure?"

Errol Morris posed that question in his 1997 film "Fast, Cheap and Out of Control," and it will be answered by this generation, one way or another. His question also reminds us that humanity is lacking in champions, the heroes and role models who reflect back into people's hearts and minds the highest and noblest ideals of life. As a result, some people see no "big issue" that should concern them, or are even aware that a human cause exists.

Doomsday off there on the horizon, you say? Well, it'll be just like another Hollywood blockbuster, special effects and all, people screaming and running around. Excuse me, I have to get over to the shopping mall before Macy's closes.

Others view human patriotism as quaint, an elitist fantasy, or contrary to national, ethnic or racial loyalties. The color of the skin or the shape of the eyes is more important than being a human.

When was the last time you heard a man, woman or child speak these words with pride and dignity? "I am a human being!"

What is your best estimate of mankind's future?

Will it "prevail" over all threats to its survival and go on to build new and better worlds?

Or is doom programmed in the genes of the human species?

Will the people of this earth "endure" for another century, or maybe even longer?

Or are they headed for the nearest exit to oblivion?

Will the nations of this world unite behind a great challenge for all of humanity, such as bringing an end at last to poverty, disease and war? How about colonizing our solar system and reaching out to the stars?

Or will political, economic, cultural and religious power struggles continue until somebody's finger presses a red button and restores cosmic silence to this sector of the universe?

Do you hear those joyous songs of love, faith and hope that rise from the depths of the human spirit?

Or are you listening to the final chorus of life sung in earth's Babylon, the lyrics of cynicism and obscenity, the tunes of loveless pleasure and mindless violence, the dirges of mourning and despair?

How high have we climbed in our own tree of knowledge?

Human destiny is controlled by how people feel about life and its place in the universe. In past eras, virtually everyone believed in a benevolent God who created the cosmos and gave His people the gift of immortal souls, a view that influenced every aspect of their thinking and behavior and inspired their brightest and darkest expectations.

In the 17th Century, Newton shifted the worldview to an impersonal, mechanistic universe, opening the way for a rapid expansion of scientific knowledge and our industrial and technological civilizations. In the 20th Century, Einstein and the quantum theorists presented us with a worldview that steadily turned ever more alien and incomprehensible, until reality faded into whatever mathematics and abstract speculations could make of it.

Cosmologists turned their attention to "other worlds of higher dimensions" and the human presence on this remote planet became merely a brief, insignificant detail in a vast, uncaring universe ruled by random chance and Heisenberg's uncertainty principle.

Is meaninglessness the final message?

Albert Einstein was "the philosopher of a new, enlightened age," according to one of his biographers; the saintly, bushy-haired genius who had solved the basic mysteries of the universe, and some people may resent any claims that might deprive them of one of their last icons.

Yet if Einstein and his successors were not fundamentally wrong about the physics of our existence, we must all, in this "enlightened age," face up to the real truth of what life and the universe are all about. A basic truth of quantum science (the Copenhagen interpretation) "says we must accept meaninglessness," concluded the late Irish physicist John Bell.

Yet, without a new vision that unites human beings in a fierce, determined mission to impress their own meaning on the meaningless and rise to the highest level they can conceive for themselves, the universe will continue its mindless cycle of spawning and destroying all forms of intelligent life.

So who cares?

A recent poll found that 51% of Americans believe that a man-made disaster will wipe out civilization sometime in the new century, and 75% said they expect a deadly new disease to appear by the year 2025. Do we care? Sure, we care, we're really concerned, but I gotta phone my stockbroker before the market closes, my wife has to get to the shopping mall before Macy's closes.

Our would-be descendants certainly won't be bothered with problems of survival if this generation ignores the question of whether humans should "endure" or fade away into oblivion. And if there were ever other forms of intelligent life flickering in the universe, they apparently didn't care either when they were given a turn in the cosmic cycling between light and darkness.

Of course, if astronomers spot a huge asteroid destined to collide with the earth at a specific time in the near future, within our own life spans, then we'll care.

If we load a computer with cold, objective facts on the ecological damage from mass consumerism and rapid population growth, the machine will tell us that human life cannot and will not pass the test of survival on this planet. But, we think, that threat is off there in the future somewhere. And whose "facts" are we talking about? Isn't this just a lot of speculation? And what about the impact on jobs and the economy if we start taking drastic measures to avert these hypothetical disasters?

Let's be realistic. Bearded prophets on street corners are always warning us to prepare for the coming doomsday, but nothing happens and life goes on. In the absence of unmistakable signs in the heavens or indisputable proof from science, we should stick to our present course and leave it to our political and business leaders to handle any problems that may come up.

Should we just give up and stop worrying?

William T. Gass, a literary critic and novelist, rated the "murderous 20th Century" with three d's: "death, dislocation and despair"—a view well supported by the dark side of our era. In a time of cynicism and disillusionment, the easiest thing would be to accept the inevitable. Humans just don't have what it takes to survive, so let's all drink up and have ourselves a merry time before the Titanic bumps into an iceberg—a millennial mood that survivors (if there are any) will remember as the *fin de siècle* of the past century.

Yet if we do care about life and its future, giving up is not an option and we have to believe that many are ready to pick up the shattered dreams of the past and go on. For now, we have only to look into the faces of the little children, and our hopes are renewed.

Should we start a new conspiracy of hope?

Walk with me down a trail that is still open to those who want to restore the importance and sacredness of life. I can't promise an easy stroll in the sun. For one thing, we'll be passing through some areas of science in our search for logical, compelling grounds for rejecting what essayist Lance Morrow calls the "absurdities" that shattered "the very mind of reason" in the 20th Century.

But that won't be the hard part of our journey. This trail is not closed to those who lack expertise in any particular fields of knowledge. The only rough going will be in climbing over the very real barriers to hope.

Along the trail we'll be looking for a fresh vision that could rekindle the human passion for life and inspire us all to give a bold, confident answer to the question facing the people of this earth in the new millennium:

Quo vadis, Humans? Where are you going now?

Part One

Journey in Time

1

The Quo Vadis Trail

> Time is one of the greatest sources of mystery to mankind. Throughout history, human beings have restlessly puzzled over time's profound yet inscrutable nature. It is a subject which has captivated poets, writers and philosophers of every generation.
>
> *The Arrow of Time: The Quest to Solve Science's Greatest Mystery*, Peter Coveney and Roger Highfield

What does the age-old mystery of time have to do with this generation's doubts about whether mankind should "endure" or welcome those oncoming waves of oblivion; whether there is something about the nature of being alive and human that could give our species a chance to "prevail" over all threats to its survival?

Those are the kinds of questions we ask when any important enterprise is up for reevaluation. Is the project still viable? Adapting to changing conditions? Achieving its primary purpose? Or does it appear to be doomed to fail, sooner or later, because of inherent defects in the design and materials?

In designing humans, evolution provided survival instincts that motivated our ancestors to work together for a common good in families, tribes and eventually nations. However, the evolutionary process trailed off and ended long before there was any need for hereditary traits that would unite the entire species behind a shared goal.

Humans have no primal instinct to preserve their species. If they did, our history books would be filled with great campaigns and crusades to protect and advance the cause of all human life—not with endless chronicles of war and conquest, not with centuries of enslaving "inferior" races, not with hopes and plans for the Armageddon. Now, as diverse peoples climb higher in the tree of knowledge and build competing technological civilizations in separate branches, we face what could become the fatal breach in our genetic heritage:

Qualities of mind and spirit that would effectively control the primitive tribal instinct that drives us to find and demonize a real or imagined enemy who seems to be causing periods of hardship or turmoil. The demon becomes the focus for growing fears and outrage, whether it's the "wicked man" down the street who must be hanged from the nearest tree, or the "evil empire" on the other side of the world that must be defeated and torn down, no matter what the cost.

In the 20th Century, the evil Hitlers, Stalins and Mao Tse-tungs used new technologies in organized aggressions that killed millions of people; in the 21st Century, many "good" men with their own righteous causes will be acquiring weapons that can kill hundreds of millions or even billions of men, women and children.

Good or evil. In terms of survival, what is the difference if either or both lead to the same outcome of death and destruction? Is the Armageddon, foreseen as the ultimate battle between good and evil, the inevitable final chapter in human history?

Is life on this planet worth defending against any threat, whether it arises from the good or the evil in human nature?

Walk with me down the Quo Vadis trail and together we'll find the "Time Key" that will unlock the gate to new directions in the search for answers. No, we won't be venturing into psychic or supernatural realms, or stepping through other dimensions of space and time. The path we'll be following is marked by this simple proposition:

The answers you seek will be found in the meaning of time.

Human destiny and the nature of time are inseparable. Can we solve the riddle of time and find some answers that could start a new conspiracy of hope? Yes, we can.

First, we have to discover the Time Key, so let's start the search by briefly retracing the footprints of those "poets, writers and philosophers of every generation" who were "captivated" by the mystery of time, according to British physicist Peter Coveney and science editor Roger Highfield. We'll be entering a forest with lots of trees, but we will pause to examine only those that mark our path to a wholly different vision of human nature and destiny.

Trust me as your trail guide, at least pending further evidence of my qualifications for exploring the Quo Vadis trail with you. I can assure you of this at the beginning of our journey:

The nature of time is no longer "inscrutable."

Starting Out

So what is time? From the early Greek philosophers, the prevailing view has been that time is a physical part of the universe and, in conjunction with space, provides the basic framework of existence. The Greeks thought of the spatial part as the "where" in which people, houses, mountains, stars and everything else existed and moved around. Time independently established the "when" in which things happened.

If we could extract time from the cosmic story, we wouldn't see anything moving and changing. If we could eliminate space, there would be no place, no setting, no stage on which the drama of existence could unfold.

Nature designed us to interact mentally and physically with the exterior world through our five senses. But the apparent framework of existence eludes our perceptions. We can't see space or reach out and touch time; they can't be tasted or smelled or heard.

So we wonder: How did Nature hang the flesh of matter and energy on that ghostly skeleton of space and time and bring it all alive—from those galaxies wheeling out there in the cosmos, down to self-aware, intelligent creatures on this planet?

That's a very broad question we'll place in the background as we examine what has eluded decades of intensive scientific research into the biological mechanisms of the brain: *the basic relationship between the mind and physical reality.* How can I make such an audacious claim? Because science hasn't found the slightest clue to the nature of time, the "profound yet inscrutable" mystery that has been hiding the origin and nature of our existence.

Every story needs a point of view from which the characters, setting and events can be portrayed as clearly as possible. Perhaps my cottonwood tree can help us start off with the right perspective in our search for the Time Key.

Why Are Trees Green?

I like to go out in the morning and sit on my porch, drink my coffee and observe the scenery of Southern California's high desert—the Joshua trees and creosote bushes, the distant mountains and the vivid blue sky.

I notice that the cottonwood tree in the front yard exhibits a "green" color that appears to be an inherent quality of the leaves; the color seems to be painted on the leaves. If I close my eyes, the green tree doesn't vanish or change its physical properties; it doesn't care whether I'm watching or not.

But with my visual perceptions interrupted, light reflected from the tree leaves is no longer impinging on color cones in my eyes and producing nerve impulses that the visual cortex of my brain processes to form a mental image of a green tree.

When I open my eyes, the image is stimulated again by the light reflected off the tree leaves, but *the tree I see* exists only in my mind; its leaves are "green" only when I'm observing them from my subjective reality of conscious awareness.

If I were color blind, I would see this scenery in shades of black and white.

> If I were blind, the world outside my mind would continue to exist in its natural state of permanent darkness.

The Dancing Skeleton

After we're born into this life, our consciousness begins to grow from a dynamic interaction between the mind and physical reality. Our brains start to assemble perceptions, thoughts and feelings into the conscious world in which we'll be living from day to day.

As we mature, our minds gradually establish what appears to be a clear separation between two "realities": our inner subjective world and an objective exterior reality with its own features and properties. However, when we analyze our self-awareness, we recognize that we contribute basic qualities to the world we perceive outside our minds.

What does separating mind and reality tell us in terms of who and what we are? Let's take the first step in restoring human worth and importance in any earthly or cosmic vision of what the future holds:

Without humans who see and think, the material world would be stripped of everything their minds contribute. It would be nothing of any significance whatsoever.

What is a sunrise if there is no eye to see, no brain to receive and interpret signals from a retina, no mind to create a glorious image of the dawning of a new day?

Nothing more than electromagnetic energy vibrating in absolute darkness.

What is the music of a Beethoven symphony or the sound of a little girl's gleeful laughter as she romps with her barking dog in the backyard, if there are no ears to hear?

Just air molecules oscillating in total silence.

So fie on scientists who tell us that the vastness of the universe makes human beings insignificant in their scheme of things. Without us, the trees, mountains, stars and galaxies would be nothing more than meaningless matter and energy, a physical reality devoid of beauty and splendor and any other mental qualities that we alone can give to it.

> *Without us, the universe could be likened to a skeleton dancing in the black void of space.*

Mind and Reality

Over the ages, the great scholars have gone off in many different directions in their pursuit of knowledge about the origin and nature of existence. Consciously or not, they were also searching for the meaning of time, which has always been the hidden wellspring for finding the meaning of life. Their quests produced a vast forest of metaphysical theories and speculations on "the causes and nature of things."

About 24 centuries ago, the Greek philosopher Plato pulled that dancing skeleton into his mind, dressed it up in the beautiful costumes of abstract logic, and announced that the exterior world—from tables and chairs to cottonwood trees and mountains—was merely an imperfect "copy" of his idealized realm of "pure reason."

His "Platonic dualism" became the first scholarly rejection of a clear distinction between mind and reality. There's the perfect original in the intellect (the "intuitive" concepts of physics theorists) and then there's the dull, boring, sometimes contemptible exterior copy that poor humans have to deal with as best they can in everyday life.

In the late 17th Century, "idealist" philosophers began wondering how we could be certain the dancing skeleton even exists in an objective reality outside our minds, since our fallible senses provide only what John Locke called "secondary evidence." They moved to the extreme of Platonic dualism and decided that nothing—no energy, matter, space or even "copies" of them—nothing existed outside their thoughts. You've heard the old notion that a tree doesn't really fall in the forest when no one is around to witness the event. Here's how George Berkeley summed up that view of the mind-reality relationship:

> All the choir of heaven and furniture of earth, in a word all those bodies which compose the mighty frame of the world, have no substance…so long as they are not actually perceived by me, or do not exist in my mind.

In the 19th Century, existentialists accepted the exterior world as an objective reality apart from humans, but then, in the words of Danish philosopher Sören Kierkegaard, agonized "in fear and trembling" over the question of whether mortal beings could find a meaning and purpose beyond the veil of their senses.

Now the more imaginative quantum cosmologists have returned to the solipsistic view of the early idealists:

> The cosmos has no existence outside their thoughts and equations. They think; therefore, the universe exists.

All interesting speculations but for now let's consider the practical side of how things work between intelligent beings and a material universe that existed long before life was given a role in the cosmic story:

The exterior world, through our senses, keeps us informed of what's happening out there, and our minds embellish the information with mental qualities—color, sound, touch, taste and smell—that help us interpret the sensory input and then react in ways that give us the best chance for surviving and going on with our lives.

Building a Consensus

In some periods of history, our ancestors projected their best ideals, secular and religious, on what they perceived as "the reality" for them and their descendants. Their laws, principles and social rules were written down in documents, such as the Magna Carta and U.S. Constitution, and in books, such as the Bible, Talmud and Koran.

As long as there was a popular consensus in society, ethical and moral principles were effective in shaping people's individual and collective thoughts and feelings and thus their outlook on life. Yes, they thought, we're all going to die (objective reality), but now while we live we must strive to fulfill our best hopes and dreams (subjective reality). Yes, humanity is still afflicted by poverty, disease and war, but now we must renew our hopes and go on to build a new and better world.

So what happened to the consensus? Do you see any signs of it in our fragmented society today? Do you hear any great orators and pundits speaking out for the cause of all human life?

> The experience of being alive is a continuous interaction between the mind and physical reality in our thoughts and feelings, and as we explore that trail we'll gain a better understanding of who and what we are and how we can build a new consensus on what life and the universe are all about.

Surrounded By Darkness

The answers we seek will emerge as we more clearly distinguish between mind and reality. Look around you now. What do you see? Walls, pictures, a rug, furniture; all various forms of matter, and your mind gives them different colors, depending on the wavelengths of light reflected from their surfaces.

What do you hear? Perhaps there's a radio playing in the background; oscillations of air molecules causing vibrations in your eardrums that your mind interprets as music or voices. Suppose you're watching television; a changing pattern of light waves from the screen that your mind assembles and deciphers as pictures.

> Now close your eyes; the world around you falls into pitch darkness, since "visible light" is visible only when there are eyes that see it. *In fact, the world in front of your eyes exists in perpetual and eternal darkness, whether you are observing it or not.*

There are trillions of other "globes of consciousness" on this planet, from insects to animals, but the world also lies before them in utter darkness. Light enters their eyes and stimulates optic nerves that send coded impulses to the brain's visual cortex, which then uses the data to create an illuminated world that exists *only in their minds.*

Light is a very narrow band in the broad spectrum of electromagnetic radiation, which ranges from radio waves to x-rays, and evolution settled on the light band as the most efficient energy for generating images in the minds of creatures with eyes. If a band of infrared or ultraviolet frequencies had been picked for human sight, the world would look different to us; but it would still be cloaked in darkness. Bats augment their limited vision by shrieking high-frequency sound waves that echo back to them and stimulate mental perceptions of prey hiding in the darkness; but waves of sound don't lift the shadows from the world either.

When you turn on the lights in your home at night, the house itself remains shrouded in immutable darkness. When you drive down a road in the dead of night, your car's headlights don't push away the darkness—except in your mind.

> Light does not light up the world; it lights up the mind. Light is the mind; dark is the physical reality. Dark is the opposite of sight; death is the opposite of life. Take away all eyes that see, and every light of the universe would be turned off. Let there be *life,* and then, and only then, was there light.

Remember those nights in childhood when "not a creature was stirring, not even a mouse"? We yearned for the dawning of a new day, when the light of the sun would make those scary shadows go away. Yet, as a fact in physics and biology, the world outside our minds remains wrapped in darkness, whether it is night or day.

> Whenever we close our eyes, we see the eternal darkness that surrounds us, whether we are awake or sleeping or dead.

Helen Keller, one of my favorite heroes in childhood, lost both her sight and hearing in infancy. Yet with the power of her mind, she used her remaining senses to establish contact with physical reality and then learn to speak, read and write. For many, she was a role model whose indomitable spirit soared above the barriers and found a special place in the experience of being alive.

Her life is also a reminder that all of us, the sighted and sightless, exist in a universe of eternal darkness, and it is only in our minds that we can build what poet Brad Leithauser hopes will be "some fortress against the dark's insidious despair."

Light of the World

Our thoughts falter when we try to distinguish between mind and reality, as they will when we differentiate between time and reality. Yet we can find previously unexplored ways to understand the differences and use the knowledge to build a fortress from which we can start a new revolution against "the dark's insidious despair."

Let me retrace the basic design:

Mental reality is the world that lights up in our minds; physical reality is a universe hidden for aeons in absolute darkness and total silence. All that exists outside of the mind are matter and energy and space. The matter, from atoms to galaxies, is moving around in complex patterns determined and controlled by four natural forces—gravity, electromagnetism and two nuclear forces at the subatomic level.

Yet, without us, the mighty cosmos is blind and deaf and dumb; it can see itself only through the eyes of life, and it can know of its existence only through the minds of living, conscious beings. So why should we exalt the universe for its beauty and splendor? It should bow down to us for giving it such nice attributes.

Life ascends to its rightful position in the cosmic story when we recognize that the glorious universe *we perceive* through our eyes and telescopes is solely a vision in our minds. Without life, there would not be one twinkle from the stars in the heavens, not one image of the beauty of a mountain range, not one whisper from the breeze caressing the trees.

> Without life, the universe is nothing more than a robotic factory that operates in everlasting darkness and endlessly churns out various forms of invisible matter and energy.

Consider the work of an artist who spreads mixtures of molecules on the surface of another substance, or the composer who dabs specks of a light-absorbing fluid on the surface of a light-reflecting solid material. So far, all they have accomplished is to arrange various configurations of molecules in patterns that match images and thoughts in their minds. It is only an audience of conscious, intelligent beings that can, in their own minds, transform those patterns of fixed or moving molecules into pictures and music, and gain a deeper sense of what it means to share in the miracle of life.

> It is conscious life that brings light to a dark world and breaks the cosmic silence, and it is intelligent life that gives existence a meaning and purpose.

In both a figurative and literal sense, human beings are the light of the world.

2

The Existential Pen

> To my mind there must be at the bottom of it all, not an equation, but an utterly simple idea. And to me that idea, when we finally discover it, will be so compelling, so inevitable, that we will say, "Oh, how beautiful. How could it have been otherwise?"
>
> Physicist John A. Wheeler

In our search for answers to the *quo vadis* question, let's bring out another little gadget for prying open secrets hidden in the mind-reality relationship: the "existential pen." According to venerable principles of logic, the pen can help us draw a line between *conceptual possibilities* in the mind and *what actually exists* in the physical reality outside the mind.

After a little practice, we'll be able to mark off more clearly the difference between our objective and subjective realities—two separate worlds that are interacting in a constant commingling of sensory input with our thoughts about what the exterior world is telling us.

When we can't tell the difference between mind and reality, we end up in a world of mass illusions where there is very little distinction between mental images and the physical reality we must know and understand if mankind is to "endure."

Strict Reality Checks

Mental images of the unknown and incomprehensible are necessarily anthropomorphic since the imagination is naked until it takes on the trappings of what we learn of ourselves and this world. For example, religious metaphors of an afterlife are based on images from this life. Heaven has pearly gates and streets paved with gold, God in the form of a bearded man sits on a throne or reaches down from the heavens, and His angels appear in a human form surrounded by halos of light.

Physics theorists who try to reach beyond the comprehensible also project ideas and perceptions from this life into their "other worlds." However, our existential pen can help us divide into two categories all of their metaphysical theories on the causes and nature of things. On one side of the line, we'll file ideas that can be understood by the human mind; on the other side, the ideas that reach beyond anything the intellect can perceive and comprehend as part of physical reality.

What is unimaginable and unknowable to the mind of reason should be reserved as the exclusive property of spiritual faith, a subjective domain where people seek a higher meaning and purpose beyond the boundaries of mortal life.

Now it's true that hunches and epiphanies that seem to transcend "conscious reasoning" can inspire ideas that may be applied successfully in some areas of human experience. But whenever the incomprehensible pops up in the stories of science, it should be subjected to strict reality checks so that we never lose the distinction between idealized mathematical models and what actually exists outside our minds.

In the words of mathematician John L. Casti:

> [Mathematics] occupies the curious position of having one foot in the real world...and one foot in the world of abstract

> mathematical objects.... *How do we know that the mathematical model of a natural system and the system itself bear any relation to each other?* [emphasis added]

As we draw our existential lines through the stories that science has been telling us, you'll be amazed—even shocked—at how often the scientific mind has projected its thoughts and images across the line separating mind from reality, and magically transformed life and the universe into the bizarre fantasies that dominated human thinking in the 20th Century.

In quantum cosmology, for example, theorists saw that mirrors reflect reverse images of this world, and then they "intuited" that every reality must be balanced by a "reverse reality" on the other side of the "cosmic mirror." Without anti- and parallel worlds that have opposite or negative features, theorists believe, the universe as a whole would lack "beauty and symmetry."

Consequently, the primary focus of multibillion-dollar research projects in recent decades has been to detect evidence of interactions or communications between pro- and antiworlds in giant "particle accelerators."

When we distinguish between mind and reality in these metaphysical speculations, we can claim our right and responsibility to make our own judgments on what's believable in science stories. We can insist that every theory must be comprehensible to the rational mind, which applies logic and good old common sense to *any* ideas, no matter how well cloaked in fancy mathematics.

> If a new or old theory of the cosmos can't be comprehended in the world community, at least in a general sense, people can make their own judgments on whether such a universe really exists, and then decide for themselves whether its "laws" should have any influence on their secular and religious beliefs.

So as we proceed down our Quo Vadis trail and spot these weird things that physicists have been saying actually exist in this universe, or in their "other worlds of higher dimensions," we'll just get out our pens and start drawing our existential lines.

Then we'll discover many logical reasons to stop glorifying the cosmos and start exalting human beings as "the light of the world."

Ask and You Shall Receive

In the category of believable ideas, we enter a new forest that we can all perceive and understand, a world where human dignity and worth will be restored and we can seek new paths for hope and meaning in the 21st Century.

How much can we learn about life and the universe in our "Quo Vadis reality"? The search for answers confronts us with the ultimate challenge:

> If the human mind is capable of asking the question, it is capable of finding the answer.

Within that broad limit, the secular realities of our existence can eventually be discovered and understood. Everything else is beyond the scope of the rational mind.

The new reality won't exclude ideas the evolving mind may be able to grasp someday, or God who requires that we know Him through the power of faith. However, if we are to survive on this planet without help from kindly angels or the wisdom that may be acquired by our would-be descendants, then political, cultural and business leaders, scientists and we ordinary citizens will all have to start paying very close attention to what that dancing skeleton out there is telling us now:

> Mother Earth gave birth to life and nurtured it for hundreds of millions of years, but now your Mom grows weary of feeding and clothing children who refuse to grow up and take care of her own needs and well-being. Shape up, you growing hordes of humanity, or ship out.

Now for the "great leap."

> It is my personal belief that we are approaching a pivotal moment in history, when our knowledge of time is about to take a great leap forward.
>
> Paul Davies, *About Time: Einstein's Unfinished Revolution*

Where Is Time?

We've sketched in some of the background scenery, so now we're ready to apply the existential pen to *time*, which appears to be half of the basic framework of existence. When we uncover its true origin and nature, we'll find the Time Key that opens the gate to an exciting new world in which life, not the universe, will be exalted.

In poetic terms, time manifests itself in countless metaphors. It's a wingèd chariot hurrying near at our backs, a bird that has but a little way to flutter, a corridor in which echo the distant footsteps of the bards. Time is a wise teacher, a kind friend, our only comforter, or it's a subtle thief of youth, the rust forming on the sharpest sword, a trumpet that heralds the beginning and the end.

On a practical level, time sorts out the mélange of information and events demanding our attention. It's a familiar guiding force or presence that imposes order in what would otherwise be a chaos of consciousness. Without time, we couldn't function as human beings.

> However, in the objective physical reality that exists without our perceptions and thoughts—in that dark material universe outside our minds, *time does not exist in any way, shape or form.*

3

A Flare in the Night

> Our present picture of physical reality, particularly in relation to the nature of time, is due for a grand shake-up—even greater, perhaps, than that which has already been provided by present-day relativity and quantum mechanics.
>
> Roger Penrose, *The Emperor's New Mind*

The nature of time has baffled philosophers, theologians and scientists over the ages and its meaning remains a fundamental mystery in modern thought. What is time? What part does it play in how the universe seems to operate? Is there a difference between the mind's perception of time and what actually exists in the world outside the mind?

The scientific view of time is "befuddling...a welter of puzzles and paradoxes," according to Australian physicist Paul Davies. A clearer picture will emerge as we use our existential pen to distinguish between the subjective world of our conscious minds and the physical reality outside our minds.

So let's start drawing some "time lines."

In the 5th Century B.C., the Greek philosopher Leucippus of Miletus reportedly became the first scholar to propose that the material universe must be a place "where everything happens out of necessity and for no other reason."

> From that classic view of reality, the universe is a cosmic machine fueled by four natural forces, whereas time is our subjective perception of the fact that the machine is in motion.

The universe is physical reality; time is in the mind that observes motion and change in that exterior reality.

Brain Clock

Nature designed her cosmic machine to operate on reliable cause-and-effect principles, which were all she needed to evolve the *non-conscious* material universe. Time (or timekeeping) didn't enter the long cosmic story until quite recently, when Nature finally got around to evolving life.

As conscious, self-aware creatures emerged from the evolutionary process, they would have had to develop some way of recognizing that things naturally happened in regular, predictable sequences of movement and change in the world outside their minds.

So that's why Nature invented the "brain clock" that gives conscious life its "sense of time." Without a built-in motion detector and timepiece in the brain, we couldn't keep track of what's happening in our lives. We wouldn't know that what goes up must come down, that when the sun rises again, it surely will set again, that the pretty woman walking down the sidewalk, one step after another, is in the process of getting from the beauty shop to her car—any of the simple or intricate segments of motion that we observe around us.

> Thus we conclude that time (like sight and our other senses) is not a component of the material universe; it is a *mental perception* whose only counterpart in physical reality is an ongoing process of motion and change driven by four natural forces.

Time is the timekeeping function of the conscious mind; the mindless, meaningless universe has no need or purpose for a separate timekeeping system of its own.

Sixth Sense of Time

Now I'm not claiming that what we call time doesn't exist in the vast world of human perceptions that stretches way beyond the space occupied by the brain. Time is everywhere; it's busy keeping track of what we experience in life. Time to get up, go to work, keep that important date, lunchtime, bedtime, a time to live, a time to die.

What I am saying again is that we, in every waking moment, *create* the world of self-awareness in which we live. We grab hold of that naked skeleton (without its own timekeeping system, it turns out to be just matter and energy dancing in the black spatial void), dress it up with our conscious perceptions and transform it into a living universe *in our minds.*

With the power of our minds and senses, we convert light into *sight*, vibrations of matter into *sounds*, smooth or rough surfaces into *touch*, compositions of molecules into *taste* and *smell*—and we perceive motion in the exterior world with our "sixth sense" of being alive: *TIME.*

> Time is the cerebral function that detects and memorizes motion in the exterior world, organizes the incoming data into a stream of mental activity, and projects it all on the mind's "inner screen."

You might think of time as the director of the story of life. You might also say that life is sort of like a dream, since the scenery, characters and script are produced by the mind from raw information delivered by that skeleton out there.

Vessels of Time

If the mind's logical faculties could operate in isolation, the world community would have no problem accepting the Quo Vadis solution to the riddle of time—you know, the "profound yet inscrutable" mystery that "human beings have restlessly puzzled over" throughout history, according to Coveney and Highfield.

However, the subjective mind tends to shy away. We can close our eyes and think—yes, in an objective sense, the green color of those tree leaves actually exists only in the mind. But when we open our eyes, the conclusion is contradicted by what our mental perceptions tell us: Those tree leaves certainly do look "green." What do you mean the sky is "blue" only when we're looking at it?

How can the world be shrouded in perpetual darkness when it's so clearly visible in front of our eyes?

And how can time exist solely as a brain function when we can see "it" everywhere, from how long it takes an adventurous baby to waddle across the living-room floor to the slow motion of stars and galaxies that astronomers observe at the outer fringes of the universe?

Why this baffling contradiction between what we see and what actually exists in the objective physical reality outside our minds?

The answers are simple, even obvious, and yet profoundly subtle and elusive because they reach to the core of how the mind works in its relationship with reality. The true nature of time rises like a flare in the night, illuminating a new and different mental landscape, but the light blinks out before the mind can fully grasp and hang onto what the eye has seen.

So there's a fascinating, if very frustrating, challenge in trying to see the new reality on that landscape. It seems the mature mind, with its deeply ingrained beliefs and loyalties, must take on a childlike innocence in order to recognize this simple, fundamental truth that has escaped the great scholars in every era of human history:

> Time is solely a creation of the mind, a basic function of the brain that establishes and coordinates the relationship between conscious life and the non-conscious material world.

In poetic terms, the brain is the universe's "Vessel of Time," the ship in which life sails over the waters of existence.

Time Is Mind, Space Is Reality

There are several ways to acquire that "innocence" and grasp the Time Key that unlocks the gate to new vistas of life's true role in the universe. Let's bring in a little historical background:

Newton thought of time as an absolute constant of Nature that regulated a "clockwork universe" with mathematical precision. Einstein rejected the classic Newtonian concept and replaced it with the radically different idea of time as a "substance" or "fourth dimension" that combines with space in the physical world outside the mind.

(Unbeknownst to Einstein, he actually took the brain's timekeeping system and tossed it across the existential line, where it landed in the cosmic ocean and magically transformed itself into a substance that blended with the "water" of space.)

In both views, time was thought to be an integral part of the universe outside the mind—an absolute constant that is somehow built into the workings of matter and energy (Newton); a fluid material that pervades the cosmos and, when combined with space, becomes the "spacetime continuum" that provides paths for planets and stars in their travels around the universe (Einstein).

Einstein stated his core idea in an essay he wrote in 1926 for the *Encyclopaedia Britannica*. It was republished with articles by other great thinkers in a special 1992 edition "celebrating 225 years of the human mind at its best." Here it is, in his own words:

> Space and time are welded together into a uniform four-dimensional continuum.

It was a remarkable idea, a radical departure from anything else "the human mind at its best" had ever come up with while pondering the nature of time. By "welding" time (mind) to space (physical reality), Einstein launched a revolution in human thinking that led to two profound and related consequences:

First, many scientists devoted their careers and vast private and public resources to exploring, experimentally and metaphysically, the bizarre trails that led off from Einstein's ideas of time and space.

Second, and more importantly, Einstein's depiction of the reality outside the mind further diminished and finally crushed the ancestral beliefs and myths that had soothed human fears of the unknown and nurtured the eternal craving for a meaning and purpose in life.

Time Is Not a Thing

In seeking the mind's fortress in a dark universe, we'll discover logical, scientifically provable grounds for the simple conclusion that time, along with sight and our other senses, exists *solely* in the mind of life. Time belongs to the mind; space belongs to the physical reality outside the mind. Time and space cannot be mixed together, except in our imaginations and mathematics.

> What we call time is the function of neurons in the brain that receives, organizes and interprets the input of sensory information—and it is that stream of perceptions and thoughts that gives the mind its sixth sense of time passing.

Time is the biological clock whose "hands" are moved by that mental stream flowing through the conscious mind (brain in operation). If there were no timekeepers, there would be no time, there would be no human beings, there would be no conscious life.

In effect, we'll be stripping our mental timekeeping system from physical reality, calling it back to its rightful and only home in the brain, and letting the skeleton (material world) use its own natural forces to keep on dancing around out there. The skeleton doesn't need our biological clocks to tell it when and how fast to perform.

Time is a timekeeping system in the brain. It is not a substance or dimension "welded" to space on the objective side of the existential line that divides conceptual possibilities from what actually exists in physical reality.

To paraphrase Bertrand Russell on the nature of electricity:

Time is not a thing like St. Paul's Cathedral.

Thinking About Thinking

It's all so simple, isn't it? And yet so infernally elusive. I'm reminded of a comment by Kierkegaard, the Danish philosopher:

> The supreme paradox of all thought is the attempt to discover something that thought cannot think.

The mind has great difficulty fathoming the nature of time because *time* is actually the *mind* that is attempting to figure out what it is. But now, with our existential pen, we can write down our first—and basic—answer to the age-old question of who and what we are:

The biology and physics of our existence are in the nature of time.

We are time. Time is us.

> There's our own "core idea" and we'll be getting into its profound implications as we continue down the trail and open gates to a wholly different vision of life and its role in the universe.

Encountering Experts

Our solution to the riddle of time can never be refuted on rational, scientific grounds, but we'll encounter critics and experts along the trail who will react first by questioning my qualifications as your trail guide. Who is this guy telling you these things about the nature of time? What gives him the right to dispute the theories of the great Einstein and his esteemed followers?

Well, it's true, I'm not a "recognized authority" in these matters. I'm just another ordinary citizen, as I assume most of my trail companions are. Yet many people like us, either consciously or in their subconscious minds, now realize they have been trading in their spiritual comforts and self-esteem for the material benefits of science and technology, and a distrust of science is beginning to spread in American, European and other cultures.

> So we won't be alone and isolated in our journey in time. There are many others, in all walks of life, who are ready to assert their right to challenge *any* theories that tell them they are nothing and their existence is meaningless.

"Grand Shake-up"

In the scientific world, Einstein's concepts of reality became the foundation on which the major theories of 20th Century physics were built. Read any physics textbook, popular science book or news report on the latest "discovery" in the subatomic or cosmic areas, and you will find his ideas of time and space woven into every theory of the physical reality outside the human mind.

The science world turned for a century on Einstein's concept of time. Time's direction (arrow) is still "science's greatest mystery." Now time's real nature is threatening what physicist Roger Penrose predicted would be a "grand shake-up" in our "present picture of physical reality." Here it is:

When it becomes known that time is not a thing "welded" to space, but *solely* a neurobiological function in the minds of humans and other conscious life forms, the theoretical framework of quantum mechanics, particle physics and cosmology will collapse, bringing about fundamental changes in the course of theories and research in the 21st Century.

4

Historic Crossroads

> The 20th Century shattered the lenses and paradigms, the very mind, of reason. The universe went from Newton's model to Einstein's, and beyond, into absurdities even more profound.
>
> *Time* magazine essayist Lance Morrow

Where is this trail leading us? Why should the world community be interested in our solution to the age-old mystery of time? Let's pause for a minute amidst all these trees and survey the forest again. On the practical side, there'll be many who'll want to know first what's the payoff for them in the *quo vadis* picture.

Well, for starters, all men and women, regardless of their lack of expertise in physics, have a personal stake in how their tax money is spent in furthering the scientific quest. If Einstein was wrong about the nature of the physical reality outside the human mind, many more billions of dollars will be wasted on fundamental research in exotic areas of the "pure" sciences.

> Resources and talents will continue to be poured into the phantom "antiworlds" and "parallel universes of higher

> dimensions" that quantum theorists and researchers construct out of empty space and *time* (i.e., their timekeeping brain functions).

But how, many people will ask, could Einstein's theories be wrong? What about $E=mc^2$ and nuclear energy and computer chips? What about all the evidence from a century of lab experiments and exploring the cosmos through telescopes?

There are alternative theories that can, as well or better, account for "the evidence," and I'll tell you about them in simple, non-technical terms in the course of our journey—a pledge I must keep if I'm to convince you of a far more important theory:

The meaning of time is the meaning of life.

Voices of Despair

Aside from research costs and other practical matters, do people have any reason to hope that current theories are fundamentally wrong about the physics of their existence? Einstein's theories still shape today's views of life and the universe, but he was also a deeply spiritual man and he believed that his vision would enhance people's feelings about their lives. His vision did indeed have a profound influence in both the cultural and scientific worlds, but not in the way he had hoped for. As he said in his later years:

> It has become appallingly clear that our technology has surpassed our humanity.

As we walk along our trail, let's listen to what other voices are saying about the worldviews shaped by science in the 20th Century. They'll help us summon up and maintain in our own minds the importance of changing course from the past and seeking new directions in the 21st

Century. Here's an excerpt from an article written by graduate student Ralph Georgy and published in the *Los Angeles Times*:

> Science has unveiled all our mystery, dissected us under a microscope and left us to live our lives in naked isolation.... Today, we find meaning and shelter in physical sensations; we escape our lives through television, the movies, music, sex and drugs. We seem to need more and different sensations to feel anything at all.

If you are concerned about crime and violence in your neighborhood, or the cultural forces now shaping the lives of your children; if you see a connection between the erosion of human self-esteem and the unraveling of social order, then I think you should take a look at *any reasonable possibility* that Einstein's theories of reality are wrong. Consider what we observe or hear almost daily about the gangs in the streets and the terrorists in the shadows. They are motivated by their own needs and beliefs, which are often violently opposed to a stable society.

Killing, they think, is the only way to attract attention, show the world how stupid and wrong everything is. If children or other innocent bystanders get in the way, let's rub them out. Their lives are worthless anyway so we're doing them a favor while advancing our own cause.

> Having given up God so as to be self-sufficient, man has lost track of his soul. He looks in vain for himself; he turns the universe upside down trying to find himself; he finds masks, and behind the masks, death.
>
> French philosopher Jacques Maritain

Time to Choose

In the context of the cosmic silence in the outer universe and its implications for the fate of life on this earth, humanity has reached another historic crossroad where it must choose between two opposing views of the reality outside the human mind:

Yes, Einstein was right; therefore, the quantum visions that arose from his concepts of reality must, in the words of astronomer Carl Sagan, continue to present humans as "an insignificant species on a remote planet whirling through space...in a vast, uncaring universe." In the face of that reality, some people have turned back to traditional religions for comfort and hope, while others have taken refuge in New Age and similar groups that create their own personal meanings in life.

To dispel the *zeitgeist* of the 20th Century, perhaps secular society should now call upon its myth-makers to take on what Einstein, in his old age, recognized as the dehumanizing "scientific and technical mentality" and transform it into a more positive view of human nature and destiny.

> When I lie dying 40, 50 or even 60 years from now, I will face death, not with fear for myself, but the burned-out, used-up world my generation and the one preceding it left behind.
>
> Christopher M. Haddad (letter to editor)

No, Einstein was wrong; therefore, all of the relativistic and quantum theories have no metaphysical relevance whatsoever to who and what we are and why we exist.

Humans have no primal instinct to preserve their species, but with the future of mankind hanging in the balance, we have to believe that people will develop the qualities of mind and spirit that will inspire them to stand up for the cause of life. Perhaps you share my doubts that we'll ever see throngs of people gathering under one flag of humanity in the real world. But then we must think of night descending on the forest of life and all human plans and hopes and dreams fading away into the shadows.

What can we do? Should we even try? Yes! Life is too precious to let it vanish into the permanent darkness that surrounds it. And it's up to us, my trail companions, to make the first efforts. Check with your friends, neighbors, business associates, the media and Internet. Occasionally, you'll hear environmentalists speaking out for the preservation of trees, fish and rare insects. But advocates of the human cause have grown silent.

5

Blaming the Stars

> The whole complex of the universe will resolve into a homogeneous fabric in which matter and energy are indistinguishable and all forms of motion from the slow wheeling of the galaxies to the wild flight of electrons become simply changes in the structure and concentration of the primordial field.
>
> Einstein biographer Lincoln Barnett (1948)

"The fault, dear Brutus, is not in our stars, but in ourselves, that we are underlings," says Cassius in Shakespeare's *Julius Caesar.*

In the 20th Century, cosmology switched the blame to the stars in the heavens, which then looked down on humans and relegated them to a role far below the rank of "underlings."

But with a new vision of the relationship between mind and reality, humans can reclaim their traditional position in this universe, for it is only through the power of their minds and senses, their hopes and dreams, that a dark skeletal world can be illuminated with light and granted a gift that only intelligent life can bestow: a universal meaning and purpose in existence.

We are time. Time is us.

Whatever our secular or religious beliefs or nonbeliefs, we can take pride in the knowledge that we, each of us as a human being, have important and vital roles in the miracle of life. The universe *we see* is not some magnificent realm with a separate, independent existence apart from us. We, individually and collectively, cloth the dark world outside our minds with light and color and sound.

We're supposed to be awestruck by the beauty and splendor of the cosmos, but let us remember that cosmic visions exist only in our minds, and without humans as witnesses, the mighty universe would be sorely lacking in admirers and champions.

In terms far more important than celestial geometry, the Planet Earth *is* at the center of the universe. After all, it wasn't a central geographical location on this planet that made New York, London, Paris, Rome and other great cities the axes on which the world turned.

Something Times Nothing

To reclaim our historic position, let's use the Time Key to unlock one answer to what Professor Casti recognized as a general problem in distinguishing between mathematical models and the reality outside the mind:

The brain's timekeeping system is a biological fact of Nature, but time can't actually leap out from our heads and fuse with empty space. Nature has been operating on her own four natural forces for many aeons, and she doesn't need or want any orders now from a "time thing" invented by human science on her Planet Earth.

> Therefore, time is not and cannot be a "substance" in the material universe outside our minds, not a structural "4th dimension" that bonds to the dimensions of space, not a "manifestation of nature's geometrical principles," as proposed by Einstein in his 1915 General Theory of Relativity.

Time, like sight and our other senses, is solely a cognitive coupling between mind and reality; real worlds cannot be constructed out of mixtures of empty space and somebody's brain functions. Logical, inevitable conclusion: Einstein's space-*time* universe and all its bizarre features exist only in the imaginations and equations of the theorists.

(Remember, I promised to tell you later about alternative theories that can explain "the evidence" that seems to support relativistic and quantum depictions of physical reality.)

Since the universe has no timekeeping system of its own, the "time factor" in conventional theories becomes zero or nothing. Something multiplied by nothing equals *nothing*, according to mathematical rules.

Lift Einstein's "time" out of the equations and we get:
(Einstein's universe x 0)+(quantum theory x 0)
+(cosmology's stories x 0)=0=0=0

Without *time* as a physical part of the universe, science's depictions of the cosmos become merely metaphors or metaphysical models that have been conjuring up fantasy worlds. Einstein's 4th dimension of time was solely a creation of his own mind; his own continuation of the "transcendentalist" thinking of the 18th and 19th centuries, a mode of inquiry in which philosophers, such as Kant, Hegel and Emerson, sought the nature of physical reality in their thoughts and intuitions, rather than by listening to what their senses were trying to tell them.

But now, in our world of existential logic where 2+2 still equals 4, we can make our own judgments on the science stories we heard in the 20th Century:

No, planets and stars are not traveling over "warps in the space-*time* continuum."

No, gravity and light are not transmitted by "ripples in the fabric of space and *time*."

No, the universe didn't explode out of a tiny speck of space and *time*, as in the "Big Bang" theory, and no, it is not now wrapped up in a "giant ball of space and *time*."

No, people and spaceships cannot travel back and forth in "*time* warps" or "*time* tunnels," except in science-fiction tales based on false notions of the nature of *time*.

> Extract Einstein's "time" from the major physics theories of the past century and all that's left is empty space.

No Einstein time means there
is no Einstein universe.
Therefore, 2+2=4.

Human thought has taken some strange twists and turns over the ages as it grappled with the mysteries of existence, but I doubt that future historians and psychologists will ever find anything weirder than the idea that time is a physical material that can be "welded" to empty space. The concept arose from exotic equations and intuitive thoughts in the mind of a very brilliant man who saw little distinction between mind and reality. But why did the world community accept his vision and all of its ramifications? How was life, which had clear meanings and purposes for our ancestors, reduced to "simply changes in the structure and concentration of the primordial field"?

I have many fond memories of Einstein, from boyhood when his work first became a major influence in my own life. But I came to believe that his theories were the primary force that "shattered the lenses and paradigms, the very mind, of reason" in the 20th Century.

Casting Out Demons

In challenging the "realities" constructed out of exotic equations, we're not knocking the many forms of math that are vital tools used by empirical science and technology to advance our civilizations. But we should always reject metaphors and metaphysical messages from equations that can only lure humanity down a path to life-degrading delusions and finally total self-annihilation.

When we turn away from what such equations have been telling us, we can start drawing some *quo vadis* conclusions of our own:

We can conclude that since Einstein was wrong about the physics of our existence, the theories and philosophies built on false concepts of physical reality are wrong—and meaningless!

We can know that we will no longer have to bow our heads and shape our feelings about life to conform with weird and logically impossible theories of science.

We can stop giving away the power of time to cosmology's bizarre universes and reclaim time as the unique creation of the mind that binds us to the non-conscious material world.

We, the people, can cast out the demons of cynicism and despair that haunted the 20th Century and find new ways to restore the importance—and sacredness—of all life.

6

The Power of Ideas

> We have gradually dissolved—deconstructed—the human being into a bundle of reflexes, impulses, neuroses, nerve endings. The great religious heresy used to be making man the measure of all things; but we have come close to making man the measure of nothing.
>
> Henry Grunwald, former *Time* magazine editor-in-chief

Whatever path we take, our personal lives and the future of humanity will be controlled by ideas of the secular and spiritual worlds in which we live. So who gave us the ideas in the 20th Century? The "recognized authorities," of course.

But they have been telling us that people are nothing and their existence is meaningless. Is that what we want to hear? That we are nothing more than talking animals, biological machines? A feeble flicker in Einstein's spacetime continuum?

Those are the "truths" that have been implanted in the human mind and many people have formed what is now a common belief system. Humans are by nature selfish and greedy, mean and miserable, violent and cruel, wicked and evil, stupid and doomed.

Now we are converting channels of Information Age technology into worldwide troughs for feeding the minds of adults and children daily with human swill. We laugh and applaud when our talk-show hosts strip off any images of dignity still clinging to individuals, institutions or human beings generally. We love the sly sex and toilet humor in our favorite TV sitcoms, and in our movies, vulgarities and obscene curses spew casually from the mouths of pretty women and handsome men. In the words of film critic Kenneth Turan, some filmmakers are now competing "obsessively" to make "even the most minimal" standards "laughable and almost quaint."

So why should mankind "endure" for another century, or even beyond our own life spans? Let's make the most of what we can get in the here and now and turn off to any more hopes that "other people" will ever be any better.

> In the past few decades, life has been degraded. No one takes it seriously anymore.... Somehow we have to turn our lives around so that drugs, despair and suicide don't become the trademarks of our generation.
>
> High school junior Kristen Perry

In the science-created realities of the 20th Century, generations of human beings could find no grand ideals that would have united them in a mission to defend the cause of life against deadly new threats to its survival. Now the entertainment and news media tell us endlessly about the corruption, greed and sleaze in human affairs and we have to turn back to the Wagners, Shakespeares and Homers for stories of great human tragedies and triumphs, for accounts of mankind's epic struggles against the forces arrayed against it, for reminders that human beings are capable of a heroic determination to achieve a magnificent destiny.

Life Is the Master

Here's the path open to us now in our journey in time:

We will continue using our Time Key to open up fundamental flaws in conventional theories of reality, then yank the flaws out of their basic foundations, watch the mighty tower of that "reality" crumble to the ground, and hear no more of the alien sermons preached from the tower's windows and parapets.

Once our trail has been cleared of those "absurdities" that shattered the "very mind of reason" in the 20th Century, we can take the first steps into a new world that offers a higher meaning and purpose in life.

The meaning of life will be found in the nature of time.

There again is the central theme I'll keep bringing up as we follow our trail through the 20th Century's wilderness of human thought and beyond. Your guide may seem to be repeating himself here and there, pointing out too many trees in the forest, but be patient; as you already know, the mind readily grasps the Time Key—and then it tends to slip out of our mental fingers. But with a little patience we can all understand—as well or better than "the experts"—the true nature of time and its power to unite the nations of the world behind the human cause.

If scholarly genius were all it took, Newton or Galileo or Aristotle (or a bearded prophet on the street corner) would have long ago solved the riddle of time, applied the solution to the basic relationship between mind and reality, and then declared this fundamental truth that reveals life as the master and the universe as the servant:

> With the power of their minds and senses, human beings are bringing light and meaning to a universe that was shrouded for aeons in absolute darkness and total silence.

The true nature of time exalts and ennobles life, and restores "man as the measure of all things." It relegates a brainless material universe to the eternal darkness in which it would exist—if there were no shelter for it in the mind of life. Indulge me if you will and take a moment to place your hands on the sides of your head. Now join me in making this declaration to the world:

> Einstein and the quantum theorists were wrong. The power of time exists nowhere in this universe, except in our minds—these temples of thought, these vessels that carry us over the waters of life. *We are time! Time is us!*

The Quo Vadis reality, and its meaning for people everywhere, will emerge as physicist John Wheeler predicted, not from "an equation, but an utterly simple idea." The Time Key will open the gate to that "pivotal moment in history" foreseen by Professor Davies and empower us to "take a great leap forward" in our understanding of life and its relationship to the universe.

> Think about it! When we cast out the absurdities that denied any meaning and purpose in human existence, we can start developing the qualities of mind and spirit that will unite the people of this earth under one banner of humanity and motivate them to work together to preserve life for themselves, for their children and grandchildren, and for all future generations.

Sure, a totalitarian regime or a powerful global business conglomerate might someday be able to bridge the gap in our genetic heritage by *forcing* a unification of our species—IF....

If it could first conquer the nations and terrorist groups that also possess the power to destroy all life—and then if it could wipe out all those in the future who would resist any form of tyranny over the human spirit—and if it could find and execute all the religious zealots who would eagerly start a fiery apocalypse in the belief that it would send their enemies to Hell and transport them to Paradise.

> Our virtual despair comes from being uprooted, from being lost in a universe where the meaning of life and of the social

> order is no longer given from on high and transmitted from the ancestors, but have to be invented and discovered and experimented with, each lonely individual for himself.
>
> Columnist Walter Lippmann (1967)

The only rational hope is in a new vision that halts the erosion of human self-esteem and inspires political, cultural and commercial leaders, writers, poets, composers and us ordinary people to take pride and joy in being members of the human family and want to place the cause of life above all other interests.

Let's walk deeper into the forest and look for more evidence of how the mind and reality work together to forge our philosophies, religions, sciences, poetry—all of our intellectual and emotional attempts to find our place and identity, individually and as a species, in the miracle of being alive.

7

A SYMBIOSIS OF REALITIES

> What then, is time? If no one asks me, I know it. If I wish to explain it to someone who asks, I know it not.
>
> Saint Augustine (354-430)

The Quo Vadis vision of hope arises from a better understanding of the nature of time, which is the basic building block in any fortress we can raise in the mind of life. But how can we convince people of the importance of time's meaning in the way they feel about life and its future?

In my experience, the problem starts with the common belief that everybody already knows what time is. So what's the big deal? Why are you taking up my time with this stuff about the meaning of time? Time is in my watch, it's in how long it takes me to get to the office or fly to New York, it's in the hours in a day or night, it's when my son will graduate, it's how old I am, it's the phases of the moon. It's...it's...well, it is what it is.

"What then, is time?"

Somebody give Augustine a pat on the back for his humble answer and we'll return to our opening premise: The simplest of questions we are capable of asking, as well as the most complex, can eventually be answered when our minds distinguish clearly between the subjective (inner self) and what actually exists in the objective (external world).

But here's the hitch: Nature designed the mind to work through a close symbiosis of mind and reality (maybe she was trying to hide the nakedness of her skeleton). Separating the two realities in our thoughts is like slicing a mouse in half and expecting the two parts to continue working as separate living beings. In a fish analogy, it's like taking away the creature's water and wondering why it doesn't keep right on swimming.

Okay, time is a basic cognitive function of the brain, not part of the brainless material universe. Yes, time does exist everywhere in the vast world of human consciousness that stretches from that hand turning these pages to the most distant part of the observable universe.

> But we must find ways to keep the *mind's time* separate from that peculiar substance, dimension, "thing" or whatever that physicists say is clinging to the spatial void.

So how do we do that? Well, we get out our existential pens, a metaphor that arose from some advice I received from an extraordinary teacher long ago and have never forgotten. It went something like this:

> Think of yourself as a tailor who's fitting cloth over a customer's body, as carefully and accurately as you can. Your design concepts take shape in the form of a handsome tuxedo, but don't confuse your ideas with the man's body. *Your ideas are not the body. The body is not your ideas.*

When an image in our minds fits some observable feature of the universe, the symbiosis of the subjective and objective is working the way Nature intended. But when we're confronted with a mystery and we start projecting our mental images onto the unknown or unknowable, without any awareness of the distinction between mind and reality, the symbiosis breaks down and we surround ourselves with all sorts of weird fantasies that we think are part of the real world.

And fantasies can be fun. Or, in terms of preserving life, they can become the deadliest form of madness.

Time and Motion

Our existential pen draws a line between the two separate but closely interacting worlds in which we live: the world inside our heads, and the world outside our heads. In the outside world, lots of things are changing and moving around. Kids chasing each other through the house, our spouse approaching for a kiss good-bye, a bird flying over our heads, cars whizzing past us on the freeway, a busy day at the office and, if we happen to be growing old, more wrinkles in that weary face we see in the bathroom mirror at the end of the day.

How do we connect our two worlds and establish a working relationship between the mind and physical reality? As intelligent beings who think and talk in languages, we attach a word or label to what we perceive as the reality and meaning of all the activity in our lives.

So we call it time. Changing time. The flow of time. The ticking of a clock, the steady energy depletion of a radioactive element, the turning of pages in a calendar—any of the devices or systems we humans *invented* to synchronize our perceptions, thoughts and plans with what's happening in the world outside our minds. In practical terms, we couldn't work together in families and societies if we didn't have clocks that told us when to get up, go to work, keep a doctor's appointment.

> "Motion and change" are general word-labels for action on the other side of the existential line, and there are a whole lot of synonyms for identifying particular forms—fluctuation, fly, jump, back up, agitation, sway, swing, alter, modify, vary, transform and so forth.

"Time" has dozens of definitions in dictionaries and most are figurative—the nick of time, time marches on, double time, part time, running out of time, Father Time, the time of our lives, a stitch in time saves nine, take your time, died before his time, and on and on.

British archaeologist Steven Mithen, who studies the cognitive evolution of the brain, has uncovered evidence that our earliest ancestors began using metaphors to connect sensory perceptions with mental images, and certainly time metaphors now fill every aspect of our everyday thinking and talking.

Metaphors of Time and Love

If I may digress for a moment to illustrate the point. Do any of you remember the "Tunnel of Love" that once was a main attraction for young couples attending county fairs? Maybe a few of them are still around. In any case, the tunnels are one way of demonstrating the universal use of metaphors in everything people think and do.

In the bygone days I remember, couples walked or rode little electric cars through the long tunnel and felt their romantic feelings stimulated by glowing images on the walls—guys and gals gazing into each other's eyes, embracing, kissing, standing in a glorious sunrise. Very few in those innocent days realized the tunnel was a metaphor of the female vagina, where love is consummated.

Similarly, we walk through the "Corridor of Time" and see images of life—birth, childhood, school days, family and career, retirement, death. Now we could say the time corridor is a metaphor of our subjective reality, where life is consummated.

Switching Labels

To establish a clear distinction between the world of "mind-time" inside our heads and that "timeless" world outside our heads, let's detach the time-label from motion in the exterior world and stick it on the neural functions in the brain that generate and process a flow of mental images of that action—and thereby give the mind its sixth sense of time passing.

(Anything else the quantum theorists see out there we'll call "Einstein's time" or "E-time.")

Now we have two labels identifying two separate worlds—the mind and physical reality, which are working together in a close partnership. A right hand, and a left hand. We might think of the two hands as conducting the symphony of life.

> But we must recognize the difference between the right hand (motion) and the left hand (time) when we talk about the origin and nature of our existence. *The time is not the motion. The motion is not the time.*

Which Came First?

In its basic form, time is a motion detector in our nervous systems, a complex of brain functions and sensory organs that enables us to perceive and memorize movement and change in the world outside our minds. We are the observers. The universe is what we are observing.

So which came first?

Was it the universe? Or the observers?

Was it motion? Or biological time systems for observing motion?

Simple questions with simple answers, yet the mind tends to go blank when it tries to look inward on itself.

Evolution designed us to project our mental qualities on the exterior world to make it real for us, a place that interacts with our needs and plans, our hopes and fears.

We see colors in the world before our eyes, and so we think that colors must exist separately from us in the exterior world. We perceive motion that matches up with our cerebral timekeeping system, and so we think that time must be an entity or invisible engine out there that is somehow driving and controlling that motion.

Let's ask the which-came-first question from a broader perspective. We draw that existential line between our two worlds—the objective physical reality on the right and the subjective mental reality on the left. On the objective side of the line, we see planets stationed around the sun, and stars in the firmament. But let's make everything completely motionless. Nothing is moving. Over on the subjective side, life has not yet appeared, so for now there is nothing there except the spatial void.

Time hasn't appeared because nothing is moving on either side of the existential line.

Now we let motion begin on the objective side. The planets, driven by gravitational and inertial forces, start orbiting the sun. The stars are drifting across the black void of interstellar space in response to gravitational (and possibly electromagnetic) forces. We move in on the third planet in the solar system and see signs of *non-conscious* life and physical activity. There, a cottonwood tree; its branches sway in a gentle breeze or thrash wildly in a storm at a rate determined by the tree's molecular structure and the velocity of the wind. A rock tumbles down a hillside at a speed and angle controlled by a combination of natural forces.

Lots of action on the side of physical reality, all determined and controlled by four natural forces.

And still time does not exist despite all this motion.

> Why not? Because the active universe began with *motion.*
> It did not begin with *time.*

Which leaves cosmology without a plot line in its stories of the origin and evolution of the universe, which takes us back to 2+2=4. Without Einstein's time, there can be no worlds exploding out of pinheads of space and time, no "black holes in the fabric of space and time," no "higher-dimensional universes" fastening on to each other through "tunnels of space and time."

On the subjective side of the existential line, we introduce the miracle of conscious life. A little girl wanders into the scene and starts observing all this action.

Now time springs into existence.

But where is time?

Time is in the little girl's mind; it is the system in her brain that connects her mind to the exterior world of action. Her eyes see sequences of relative positions—the changing angles of the tree branches shaking in the wind, the arc traced in the air by a birdie flying overhead, the little doggie dashing out from the gate and coming closer and closer, and her mind memorizes the sequences of motion—an ongoing mental process that gives her the sixth sense of time passing.

Time and the little girl's conscious mind exist only on the subjective side of the existential line.

The universe of matter, energy and space exists only on the objective side—in its natural state of eternal darkness and silence.

Number Mysticism

Time flows through the mind; it's not a river running through the universe, except in the imaginations of scientists who feel driven to create worlds that match up with their equations. Without a time river, relativistic and quantum ships just can't sail anywhere, except deeper into fantasy lands. Meanwhile, Nature will keep on doing her own thing without any help from their equations.

In 1918, German philosopher and historian Oswald Spengler contended that when scientists start reducing natural phenomena to mathematical formulas, their work tends to degenerate into a kind of number mysticism.

Since then, math wizards have taken on the role of the Sorcerer's Apprentice in conjuring up the bizarre worlds in which people of the 20th Century thought they were living and dying.

8

Human Consciousness

> The important thing in science is not so much to discover new facts but to find new ways of thinking about them.
>
> Physicist William Lawrence Bragg

The nature of time also gives us new insights into what has long been another "something that thought cannot think": the source and nature of human consciousness. Suppose you're sleeping now. The *conscious* part of your brain is closed down for the night, so your self-awareness is turned off (except maybe for some dreams floating around in the unconscious part of your brain). Then your ears send you a strong signal that causes electrochemical energy to start flashing over a network of neurons. Your eyes open and start sending in a stream of visual perceptions of your bedroom. Memories from the past interpret your perceptions and tell you: My wake-up alarm says it's *time* for me to get up and start a new day. So it is your brain's biological timekeeping system that has switched on your awareness of being alive.

> Time is the creator of consciousness! When you're conscious, you are keeping time. When you're keeping time, you are conscious. If you could look into a mental mirror, your con-

> scious mind would see itself as a flow of perceptions, thoughts and feelings. Whenever the time flow stops, the mind becomes unconscious.

Intelligent life, which bestows splendor and meaning on the universe, *is* the constantly modulated electrochemical energy racing over memory tracks in the brain and feeding signals to the nervous system to control conscious and unconscious functions and movements of the body.

Does that mean we are part and parcel of time, and vice versa, just as we are part and parcel of everything else in nature, and vice versa? No. No way. Those connections can only circle back to the old world of transcendentalism, where time again becomes a "something" with a life of its own, independent of our lives. Then time takes on its familiar metaphorical forms, such as a mighty river on which our little boats sail briefly, and when we sink below the waves, the river just keeps rolling along. No way. When we die, our time dies with us. If all conscious life ended, all time would end. Time exists only as the neurobiological structure and energy at the core of our being.

We could not exist without time.

Time could not exist without us.

We are time. Time is us.

What Then, Is Pain?

Our conscious experience of pain can also help us distinguish between mind and reality, and thus further clarify the picture for those who still don't see the crumbling of science-created worlds all around them.

Suppose your dentist is drilling out a cavity in one of your back teeth—but without first injecting a pain-killing drug. So *where* is the pain you are feeling now? It's certainly not in the dentist's hands or instruments. Nerves in your jaw are sending urgent signals to your brain, which your mind interprets as the pain now dominating your conscious awareness. You generate your own pain.

Pain is you. You are pain.

Pain isn't "welded" to the physical universe outside your mind. The sun isn't howling and groaning from that terrific heat generated in its innards by nuclear fusion. A tree doesn't cry out in pain when a woodsman chops it down with his axe. Pain is solely a property of humans and other self-aware life forms; we know pain doesn't exist on the other side of the existential line.

If the dentist anesthetizes the nerves in your tooth and jaw, you would feel no pain from his instruments. Pain would cease to exist (at least for you).

Similarly, time isn't "welded" to space. Time is in your mind, where it's keeping track of what's happening in your life. You generate your own time; it exists only on your side of the existential line.

Time is you. You are time.

When you doze off, time ceases to exist (at least for you). When you awaken and become conscious again, time resumes its natural function of memorizing and processing sensory information from the exterior world and guiding a stream of thoughts through your mind, giving you the sixth sense of time passing.

So let's cross out pain, sight, sound, touch, smell, taste *and time* as physical properties of the universe outside our minds. Then we can brush aside these weird, life-degrading stories that begin with the notion that time is a substance or dimension embedded in space on the other side of the existential line dividing mind from physical reality.

> We are time, and we are joy and sadness, love and hate, hope and despair, and we are all the other properties of intelligent life that exist solely in our minds.

Playing Nature

Nature designed the brain to handle a very rapid flow of information from five senses (especially visual), and let's call her product the basic motion-detection system she needed to connect conscious life to her non-conscious universe of movement and change. The connection just wouldn't work without a brain system that matched swift sequences of sensory data with how fast things were actually happening in the world outside the mind.

Imagine taking over Nature's job. After aeons of hard work, you have evolved a material universe of planets, stars and galaxies. Now you select a planet where conditions are most favorable to evolving your carbon atoms into forms of life. To connect *conscious* life to the material world, you give your creatures eyes and ears and other sensory organs.

But that's only half of what you'll need to get conscious creatures to interact with the world outside their minds. You have to evolve a brain system that *memorizes* sensory information and organizes it into a stream of perceptions and thoughts.

Look Out for Boulders

Say one of your early creatures is out hunting for food and spots a big boulder on the side of a steep hill. Is it stuck there? Or is it rolling toward him? Well, there's only one way he can know the difference. If the boulder is not moving, his mind will *memorize* an unchanging sequence of visual perceptions, which tells him not to worry about the boulder.

If the boulder is rolling toward him, he will perceive and memorize it in one position, then relate that memory to its next changed position, and then the next and the next and so forth. It would be sort of like frames of film running through a movie projector and giving a viewer an impression of continuous motion.

If your guy is going to survive an approaching boulder, that ongoing comparison of immediate perceptions with a lengthening string of past memories had better tell him to get out of the way.

As your creatures deal with ever more complicated survival situations and evolve into intelligent beings, the motion sensor in their brains develops a complex process of perceiving movement and thinking about its meaning for them—like boulders and saber-toothed tigers heading in their direction—and eventually they stick a word-label on their motion-detection brain functions: TIME.

Isn't that a plain, indisputable fact of our existence as conscious, thinking beings? Without an efficient time hookup between mind and reality, we would get only jumbled, confusing perceptions of what was going on in our lives. We would be afflicted with a rare disease that psychologists call "motion blindness." A boulder (or car) would run over us whenever we wandered into its path.

The Great Connection

Nature evolved the time function for a single purpose: connect life to non-life. All she needed at the beginning was a basic motion-detection system hooked to a primitive brain that memorized and interpreted sequences of signals from the world outside the mind.

> Over hundreds of millions of years, the brain's motion sensor evolved through higher and higher levels of complexity and finally produced conscious, thinking beings *who are themselves a sophisticated, biological time system.*

All forms of conscious life, from insects to birds to humans, emerged from the eternal darkness of existence as they evolved motion detectors and other senses in their brains. Each species' intelligence and self-awareness is proportional to how much knowledge can be stored in its brain, and its survival is determined by how efficiently the brain's mechanisms can utilize perceptions, memories and instincts to interpret what's moving and changing outside the mind.

By the way, are you aware that the "color" of a person's skin exists *only in people's minds*? All human skin is the same except for minor differences in surface molecular configurations. Your skin seems to be black in the minds of others if it *absorbs* all wavelengths of visible light. Your skin appears to be white if it *reflects* all wavelengths of visible light.

Absorb or reflect. A trivial, insignificant difference in all that makes up a human being.

All humans are Vessels of Time sailing through the universe in search of their destinies.

9

Nature's Clock

> Nothing troubles me more than time and space, and yet nothing troubles me less, as I never think about them.
>
> English essayist Charles Lamb

Are we all confident now that we can explain to our friends and associates the true nature of time and how it can be a rallying cry for those who want to stand up for the cause of life? Will they want to join us in the shelter of a new fortress?

Well, you're right. It's hard to grab hold of time and lift it up for the inspection of others before it slips through their mental fingers too, and reappears in a thousand different places, daring us to try another way of explaining that "profound yet inscrutable" mystery.

Let's return again to Augustine's question—"What then, is time?"—and answer it by identifying and defining time in three basic forms or states.

FIRST, and most important: The conscious mind is the universe's one and only system for keeping time. *We are Nature's clock.* So far as we know in the secular view of the world, there's nobody or anything else that has the human mind's power to perceive, remember and think about scenes of Nature in action—and thus become conscious.

Our time system coordinates our perceptions and thoughts with what's happening around us, and it works fine in the subjective reality of our everyday lives. But the brain doesn't keep precise time, since it must rely on its fallible, imperfect memory for preserving records.

SECOND, time is all the gadgets and methods we humans invented for keeping more accurate records. Our early ancestors started with marks and drawings on cave walls that connected their lives to the perceived motion of the sun and stars, the phases of the moon, the changing seasons of the weather. Eventually, they *invented* hourglasses and pendulums, and today we have clocks and complex devices that we *invented* to divide our time into seconds, minutes, hours.

For long-term records, we *invented* calendars that divide our time into days, weeks, months, years, centuries, millennia. To predict and calculate motion, we *invented* mathematics that uses the letter "t" as a symbol of our most precise timekeeping systems.

AND THIRD, time is the countless metaphors that language-thinking humans use to interpret signals of what's happening in their lives. Hey, you've been working hard all this *time*, so now it's *time* to take a vacation in Tahiti, *time* to bed down in the hotel and make love with your spouse for a long *time*, then set the alarm so you'll wake up in *time* to catch the native festival in the village, scheduled to begin at 8 p.m. local *time*.

> The brain's timekeeping system is a neurobiological fact of Nature. Time is the perceptions, devices and metaphors we use to organize and run our lives.

Omnipresent time arises from the brain's structure and energy, and from the mind's capacity to memorize and retrace sequences of sensory information, organize the input into a stream of thoughts and feelings, and thus acquire a sense of time flowing from the past into the present and on into the future.

Time means we are alive.

We are alive means we are keeping time.

We are time. Time is us.

Nature's "Time"

In a strictly metaphorical sense, we could say that Nature has her own timekeeping systems. Cut down a redwood tree and count the number of rings in the stump, and that's one measure of the "seasons of time" that have passed since the tree grew out of its seeds. But the tree doesn't know that it's "keeping time." It doesn't give a rip how long it takes to grow up.

A steady deterioration of molecular structures, hastened now by the blazing desert sun, is producing physical changes in that Mercury Cougar parked in my backyard. But when I climb in the driver's seat and start up the engine, the machine doesn't tell me: "Hey, I can't keep hauling you around much longer. My time is nearly up!"

Mountain ranges are born in volcanic eruptions or massive earthquakes, and "over time" the wind and the rain slowly erode the peaks and slopes and, given enough "time," they make the mountains disappear. In a solar system, planets orbit a sun for a large but finite "number of times" until finally the system collapses and changes to another form of matter and energy.

Galaxies wheel in the cosmos and their states and relative positions are one measure of the "universe's time." Stars explode in the heavens, but they don't ask each other if they just happened to blow up "simultaneously," or at different "times" measured in "microseconds" or "billions of years."

> Nature has no sense of time, no clocks and calendars of her own for scheduling her operations. She's not saying to us or

> anyone else who may be listening: "Hey, I'm not an old lady yet. I've been around for only 15 billion years! So give me some more time to show you what I can really do."

When we distinguish between metaphors and the physical reality outside our minds, we recognize that rings in tree stumps, the motion of planets and stars, and all other phenomena have no purpose in the non-conscious material universe other than to follow the rules of four natural forces—in absolute darkness and total silence.

Natural phenomena acquire a meaning only in the minds of conscious, thinking beings who observe and connect them with their sixth sense of time passing.

10

FAITH IN HUMANS

The meaning of time exalts and ennobles life.

The meaning of time, which emerges from a clearer distinction between mind and reality, is most easily applied to our lives in generalized terms. But now let me try the personal approach, and I would ask you to forgive me if I wax sentimental; perhaps your own reflections will be stirred by memories from the past, when you also desperately needed "some fortress against the dark's insidious despair." So here goes.

I am time. Time is me.

I like to go for early morning walks in Southern California's high desert, and so does my border collie "Friend," a name we agreed on when I picked her out from dozens of other dogs at an animal shelter and she wrapped her little puppy's body around my ankles to show that she also needed a companion and friend. We walk together for miles and along the way my spirits are lifted by inhaling the clear, cool air and by observing the tall, dark-green creosote bushes, the wild, strange Joshua trees with their cones of leaves and seed pods jutting out of the ends of the branches, and there's the magnificent San Gabriel mountain range defining an irregular line at the bottom edge of the vast blue sky.

Friend, now full grown, perks up her ears when birds chirp in the trees, barks excitedly when a rabbit occasionally appears in the distance, and she sprays drops of urine in the weeds here and there so that on future trips her sniffing nose will tell her that this is her favorite trail too.

Then I think: All of this natural splendor exists only in the minds of Friend and me and the rabbits and the birds. It is in the mind of life that these lovely pictures are painted with light and color. Without any conscious life, this beauty would be shrouded in darkness and silence. We can't even imagine a world in which there is no one to see and know it.

Wo Sind Sie Hin?

In such moments, I am reminded of how precious all life is. I'm also reminded now of a verse from a Heine poem that I first heard from a lovely, 16-year-old girl, long ago. We met in a city that reeked of death and destruction, and after she had lifted the fogs of hatred and despair from my eyes, we spent our last hours together, wandering hand-in-hand in the deserted, moonlit ruins of war. She sang *O Holy Night,* in French as she had learned it from a friend, in a slow, waltz tempo, as she danced in my arms, amidst the rubble of what had once been a grand cathedral. Her high, clear voice trailed off into the silence of the dead city, and then she whispered to me, "Remember, no more sorrow...although I must say...your tears do fall on my face...like sweet rain...."

Here's one translation of Heine's verse:

> I have known many a beautiful child,
> And I have had many a good comrade.
> *Wo sind Sie hin?* Where are they now?
> The wind blows, the waves foam and wander.

Heine died more than 140 years ago, leaving us his thoughts on the value and importance of human beings in their worlds of time. He reminds us that life, in the present and the future, becomes even more precious when we think about the people we have known and loved in the past. *Wo sind Sie hin?* Where are they now? The wind blows, the waves foam and wander.

Crowning Achievement

What we call time is consciousness in motion, a stream of recorded perceptions retraced in the brain, a progression of thoughts and feelings from energy flashing over networks of memories. Time is the cerebral engine that causes and drives a continuous, ongoing consciousness in the mind. Time is the bridge between the brain and self-awareness, the neurobiological structure that bonds conscious life to the non-conscious universe.

Time and consciousness are inseparable. Without time, there would be no consciousness. Without consciousness, there would be no time. Time is life. Life is time.

Consciousness spreads time throughout the mind's world of subjective reality, even as the brain's motion-detection system propagates consciousness throughout that vast reality.

Cogito ergo sum, Descartes concluded. I think, therefore I am. When we follow his line of thinking deeper into the core of our being, we reach our own intriguing conclusion:

> Time and consciousness are mirror images of each other projected out from the mind. Working together, they manifest the crowning achievement of the universe: the evolution of matter and energy into conscious, thinking beings who see a world radiant with light.

What Is a Human?

Who are we? Where did we come from? *Quo vadis?* Where are we going?

Questions that have fascinated and baffled the mind since the dawn of human consciousness. Why does existence exist? Did it appear out of nonexistence? But how could nonexistence exist? Why should we think that our lives must have a meaning? Who or what could provide a higher purpose for life?

How can the mind ask itself such questions and supply its own answers? It's like asking that person in your bathroom mirror to solve the puzzles of life, tell you what it all means. But the other person speaks only when you speak, scowls only when you scowl; the image in the glass, like time and consciousness, exists only in your mind.

Yet by using our existential pen, we can connect the enigmas of human existence to the nature of time and draw inspiring new pictures of the future of life. We can revive the "One World" vision that motivated advocates of the human cause in the wake of the Second World War, when national rivalries and new technologies began to write what many still see as the final chapter of life on this earth.

> That's what our journey is all about. Solve the mystery of time and we can better understand the perplexities and great possibilities of life. We can take pride in being human and members of the human family, and then we, the people, can renew the long, hard campaign to unite diverse nations and cultures under one banner of humanity.

Let's continue walking down our trail, and after taking some more looks at how the "scientific and technical mentality" has given us distorted pictures of life and the universe, we'll reach the high ground where we will see the "Quo Vadis Vision" of the future.

Along the trail, we'll discover more of what thought has not been able to think, for now we know that we can acquire deeper insights into the nature of our existence as we learn more about the cognitive coupling between mind and reality.

We won't solve all of the mysteries by any means, but in a hopeful, refreshing vision of humanity's future, we can believe that our descendants will find many more answers to any *quo vadis* questions the mind is capable of asking the knowable universe, a continuing search that will elevate our species to the highest and noblest destiny that humans can conceive for themselves.

Religion of Life

Some years ago the *Los Angeles Times* published my interview with Leonard Wibberley, a prodigious author, fiery Irish expatriate and life-affirming philosopher. *The Mouse That Roared* was among more than a hundred books he wrote in a creative outpouring that would exhaust a battalion of other writers.

"Man's religion is life," he said, "and his purpose is to continue life." He was moved by archaeological evidence that brutish Neanderthal Man had felt tenderness and love for his dead. "We owe him. He endured as we must endure."

Mr. Wibberley, who died about a year later, said he had no expectations or desire for personal immortality. He would not be looking for a star to guide him through the darkest of nights when he closed his eyes for the last time. But he believed in the immortality of life, and "I would be bloody-well glad if my little atom of being is part of intelligent life ascending, if only by millimeters over the aeons, toward perfection and truth."

Like some of you, I'm one of those unbelievers who can never give up searching for the paths of hope that we walked in our youth; for a bridge between what reason tells us and what faith may reveal.

However, it seems to me, faith in God should be matched by faith in human beings, a belief that despite all of our sins and follies, we are worthy of salvation by a higher being, deserving of His gift of an immortal soul, that we are more than biological machines, that we are more than quantum fluctuations in some alien primordial field.

Human beings are time, and they are light and sound, touch and scents and taste, and they are love and compassion, joy and beauty, hope and faith, and they are visions of the heavens and the earth and they are all the other qualities of the human mind that give each of them a vital role in the miracle of life.

Part Two

Searching for Reality

11

Einstein's Time

> Einstein penetrated the realities of nature more deeply than any man before him and changed not only our concept of the universe, but our everyday life…[he is] the philosopher of a new, enlightened age, and an astonishing tomorrow.
>
> Lincoln Barnett, *The Universe and Dr. Einstein*

Why did Einstein want to combine space and time into one substance—his spacetime continuum where planets and stars create the "warps and curves" along which they travel? The combination can't be proved empirically because his space and time are invisible substances or entities in the distant cosmos where an actual physical mating of the two elements can't be witnessed from earth.

But neither can we prove a negative—that the spacetime continuum doesn't really exist—just as we can't prove that people never have out-of-body experiences or never see UFOs in the skies. So we're expected to believe that exotic equations, supported by limited empirical evidence, are proof that planets and stars do, in fact, force space to coalesce with time so they'll have pathways for motion.

We can "see" the space and time that our cars move through on trips to the supermarket, and while the two elements are obviously connected in our minds, we don't blend them together in one physical, flowing substance to carry us to where we're going. (That reality would be good for boat builders, bad for automobile makers.) In the relatively small reality we experience on earth, we know that space is not actually combined with time in some physical way outside our minds.

So we wonder: Did Einstein discover a world "out there" that our limited minds simply can't comprehend? Or have the theorists merely constructed their own universe out of "higher dimensions of space and *time*" (i.e., their motion-detection brain functions)?

Let's dip back into history again.

Intuitive Thinking

In pre-Newtonian eras, our ancestors also craved answers to the mysteries of existence, but they had few objective explanations for how the world operated. So they used that innate human ability to project thoughts, feelings and needs on the exterior world. By merging what mind and reality were telling them, they came to believe that supernatural forces must be controlling everything from the perceived motion of the sun, moon and stars, down to weather conditions on earth.

That's called intuitive thinking, "the ability to perceive and know things without conscious reasoning"—a capacity of the mind that often takes over when not enough objective knowledge is available to piece puzzles together.

In the 17th Century, Isaac Newton replaced the old gods with his objective view that gravitational forces govern the universe, and he devised a mathematical system that accurately predicts and quantifies what's happening in the material world. His laws of gravity and motion became the basic tools of science and engineering—and still are today in all but Einstein's relativistic realms.

In the 19th Century, new phenomena and long-standing puzzles, such as a very slight deviation from Newton's equations in the planet Mercury's orbit around the sun, began to cast doubts on the accuracy of his laws under all possible conditions.

The mind inevitably connects *rates* of motion with *time*, but nobody knew what time really meant. It seemed to exist everywhere, but "what then, is time?"

Cloaked With Math

Einstein tackled the ancient riddle, believing that his interpretation of time's nature would solve old and new mysteries confronting science at the end of the 19th Century. (Underlying his basic theories is the age-old metaphor of time as a great river flowing to infinity and eternity.)

In seeking his answers, Einstein essentially returned to the ancient mode of thinking and cloaked it in exotic mathematics that he and his associates developed. As he contended in one of his early papers:

> It should be possible to obtain by pure deduction the description…of every natural process, including those of life.

His approach to solving mysteries is called "deductive" or "intuitive" reasoning, a way of thinking that has replaced the classic inductive method in major theoretical and metaphysical areas of the physical sciences.

Einstein "welded" time to space. In effect, he attached the fundamental mystery of human consciousness to the fundamental mysteries of the universe.

If he had connected time to the conscious mind, scientific and cultural histories would have taken a much different course—perhaps more like the Quo Vadis trail we're following now a century later.

Tests for Time

There is no doubt that a timekeeping function exists in the brain; even on the most obvious, practical level, we would all be in big trouble if that motion sensor could be snatched out of our skulls. Without it, we wouldn't be able to see and memorize the movement of things in front of our eyes; we and our world would be "frozen in time." Eventually, neuroscientists will prove, down to the finest details of neural locations and interactions, that time is indeed an indispensable form or "partner" of the conscious mind.

Can physicists prove that Einstein's time exists in the brainless material universe outside the mind of life? If they used the empirical approach, researchers could try to extract a sample of his "time substance" and put it in a test tube or under a microscope. Or maybe pluck "the thing" out of space and give us a look at it on the evening TV news and in morning newspapers.

However, the project would distract them from their research into "cosmic superstrings" and "parallel universes of higher dimensions," so perhaps they would offer their equations again as "proof" that Einstein's time really does pervade the cosmos; that it is not merely a projection of mind-time on the universe's natural forces.

Equations don't need time (symbolized as "t") if nothing is moving or changing. You can count the apples in one typical tree and then calculate the total apples in the whole orchard, but if you want to figure out how fast an apple falls to the ground, you have to build the time factor into your equations.

$E=mc^2$ relates energy (E) to mass (m) and has time (t) in its proportionality factor (c^2)—the square of the speed of light measured in miles or kilometers *per second.* So in any equation that calculates motion, a clock or other timepiece must start feeding numbers into the time factor before the math can take over.

But who invented the clock and math?

Humans, the creators of time.

Off to the Cosmos

In devising a form of geometry that would fit his special vision of reality, Einstein intuited that the time dimension in his equations must be matched by a physical dimension out there in the cosmos.

But what is it? Is his cosmic "time dimension" like a stick made of some unknown type of matter or energy? Does Nature employ a secret "fifth force" to hook up his time dimension with the three-dimensional sticks of space, and then use the four-dimensional "vehicle" to transport planets and stars around in her universe?

In some aspects of relativity, Einstein's time becomes sort of a clock-like force or "entity" that, in later decades, the "Big Bang" theory created out of "nothing" at the beginning of the universe, and ever since then E-time has been the overseer of cosmic operations. Without E-time, cosmology's universe could never have started out, and Einstein's would have collapsed if he hadn't invented a "cosmological constant woven into the fabric of space and *time*" to maintain things in a "steady state."

In our Quo Vadis reality, the *active* universe, if it ever had a beginning, operates solely on its own four natural powers—gravity, electromagnetism and two nuclear forces—and without any help from our time dimension.

> The mind of life is the one and only home for time. The brainless material universe hasn't the slightest clue to what time is all about.

Meaningless Time?

Einstein also thought of time in still another form: a fluid material that flows like a river—but at variable speeds and even the reverse direction—when it is hypothetically transporting spaceships or other objects at near the speed of light (relative to an observer).

Whatever it is, E-time can claim no separate existence and meaning of its own. It must be blended first with space (another substance?); otherwise, according to Einstein, his time is "absolutely meaningless."

If meaning is ever lost, I guess E-time, in all its forms, would have to vanish out of existence. But would Einstein's *space* become meaningless too, if left all to itself in the cosmos? Would his space, along with the planets and stars presently inhabiting it, have to join "the thing" in meaningless nonexistence?

> Intuitive reasoning can be rather like a magician's sleight-of-hand tricks. He shows us a coin in his left hand, but when we point at where we last saw it, he makes the coin vanish and reappear in his right hand—and finally disappear from both hands.

Dazzled by his show of such amazing powers, we give him some coins of our own that he can put in his pocket.

Bump and Grind

The "logical completeness" of his equations, Einstein once remarked, required that if "a single one of the conclusions" drawn from them proved wrong, the "whole structure" of relativity "must be given up." So far as I recall from reading his work, he never considered the possibility that the whole structure of his theories had been built on distortions of the mind-reality relationship.

Einstein's relativity equations do match up beautifully with several observed features of the universe. That's what good equations are supposed to do. Would you expect a very brilliant, dedicated man to work for more than a decade on his equations—and then they didn't fit anything?

But mathematics, in itself, doesn't prove anything. Equations can be manipulated to fit any real or imagined feature of the universe, and if contradictions arise between math models and empirical evidence, we can conjure up "other worlds of higher dimensions" where our equations seem to match up better. As Nobel laureate Leon Lederman puts it:

> You manipulate the equations, bump and grind and, sooner or later, out comes the answer you're looking for.

Equations, instruments and computers can measure the movement of planets, stars, galaxies and baseballs, but motion in itself is not empirical evidence that E-time is somehow embedded in the spatial void. Simply *measuring* the flight of a baseball isn't what proves its existence; we confirm its reality by looking at it, catching it in our hands, touching it. (But we should be wary of calling on the psychic or intuitive if a baseball's flight or other feature of reality seems to change from what we have observed in the past.)

Don't Freeze Up

In a metaphorical sense, Newton saw a universe operating with "clockwork precision," but he didn't mean that some sort of clock was actually controlling the motion of planets, stars and apples falling from trees. If he had discovered our Time Key first, he probably would have concluded that time is part of a mathematical system that humans invented to predict and calculate motion driven by gravitational forces.

Built-in forces (from a coiled spring or electric motor) make that clock on your wall tick at a dependable rate, and the position of its hands can help you coordinate your daily activities. But natural forces in your mind and body—not the clock itself—are regulating what you're doing. If the clock stopped ticking, you wouldn't freeze up, would you?

Clocks, calendars, phases of the moon *measure* time.

Humans *create* the time that's being measured.

Motion is determined and controlled by gravity and three other natural forces, and through our thoughts and equations and instruments, we project our sixth sense of time on motion so that we can keep track of what that skeleton (or our spouse or kids) is doing.

> Physicists are understandably reluctant to abandon the theories of physical reality developed in nearly a century of dedicated study and extensive research. One way to resolve the problem and get on with the vital scientific quest would be for government agencies or other funders of current research to offer a million-dollar reward to any physicist who could prove that E-time actually exists as a clock-like force, dimension or flowing substance in the objective reality outside their minds and equations.

Meanwhile, let's proceed down our trail with the theory that time, like beauty, is in the eye of the beholder.

12

Geometry Is Reality?

> The imagery allows us to move forward more rapidly, but the truth is in the math.
>
> Physicist Kip Thorne

I've been saving this subject until we had walked past some of the basic scenery and characters along the Quo Vadis trail, and now I think we're ready to take on concepts of dimensions that play a critical role in the stories of cosmology and quantum physics.

The brain's timekeeping system enables us to predict and calculate motion, so time became a "dimension," a yardstick for measuring the flow of events we observe or imagine. We use the three dimensions of space (length, width, height) to plot motion—for example, the flight of an airliner on a trip from Los Angeles to New York. Any point on the curve represents one position of the airliner, and the summation of the points covers the entire flight. Simple Euclidean geometry.

If we want to determine the position of the airliner at any time (sequence of positions) and predict when it will arrive in New York, we link each "stationary point" on the curve with a "time point" specified by flight controllers or attendants. That's motion in the world of ordinary experience. However, the intuitions of theorists perceive the movement of airplanes and other things in a quite different reality.

In their reality, the dimensions of space and time are measuring sticks that can be lifted out of their thoughts and equations and assembled in various configurations that are then believed to have a separate, independent existence outside of their minds.

The tuxedo is the universe. The universe is the tuxedo.

Mathematical Sticks

Dimensions, like time, sight and our other senses, are creations of the human mind. They are part of a geometrical system we *invented* to specify the size and shape of objects, such as a box, and the box's existence and physical properties don't depend on how we perceive and measure it. Yardsticks and clocks are just more hunks of matter until we give them a purpose, which is to use *our* dimensions of space and time to define objects and how fast they're moving.

The blurring between mind and reality begins when theorists think that things, from subatomic particles to galaxies, couldn't exist without their mathematical sticks to define and somehow hold them together in whatever shape or form they observe or imagine.

> So they detach the sticks, add a few more "other dimensions of space and time," then use their collection of dimensional toothpicks to construct the worlds of cosmology. They magically transform the mind into physical reality and ultimately leave humans stranded in an alien world where nothing multiplied by something equals *nothing*.

Elevating Humans

Mathematics, which is all about the relationships among dimensions and physical properties, defines many observable features of the material universe with beautiful precision, but its actual connection to reality does occasionally stir controversy in the scientific community.

British mathematician G. H. Hardy and most other theoreticians believe that a "mathematical reality" exists separately in its own realm and provides the laws and principles that govern how the universe operates. However, from our existential perspective, we must take the side of a small number of scientists, which includes mathematician Reuben Hersh and neuroscientist Stanislas Dehaene, who contend that mathematics has no existence outside human minds.

They say that numbers, symbols and math principles are inventions of the mind that we project on the material universe in our efforts to understand it. From that "humanist" viewpoint, the role of human beings is elevated, for they alone can sit in judgment on the cosmos and decide whether it is "rational" by their standards.

(If Nature chose to start off her universe by exploding it out of an infinitesimal speck of space and *time*, I would say she is the most irrational, unreasonable woman I've ever known. Well, almost.)

Over the aeons, matter could take on every possible configuration, including an absolutely "perfect circle" whose area exactly fits our idealized equation, *pi* multiplied by the square of the radius. But such circles (with a pancake's width to give them some room for real matter) would form as a result of natural forces and conditions, not in obedience to orders sent down from a higher mathematical reality.

> The universe of motion came first in physical reality; the laws and principles—absolutely meaningless without *our time*—came second in the minds of humans when they started correlating their inner subjective world with the universe they perceived outside their minds.

Why in the devil should we give that dancing skeleton the credit for figuring out his choreography? The next thing we know somebody will nominate him for a Pulitzer Prize in the performing arts.

(Nature is usually personalized as "she." So I thought maybe I should refer to her naked skeleton as "he.")

Confusing the Gods

In the example of the airliner's flight, the time "dimension" is a required factor for motion in equations, and so theorists think their geometry must somehow control the plane's movement over the "fabric of space and time." Moreover, the equations work equally well in the forward and backward directions, which they think means the airplane can also "fly back in time" (that is, according to Einstein, if the flight is witnessed by an observer who happens to be traveling in the vicinity at near the speed of light, relative to the airplane).

So what the theorists propose is a four-dimensional map on which the entirety of space and time is laid out. And since time doesn't exist point by point on the map, but all at once, they think the past, present and future must all be going on *simultaneously*—an unimaginable situation that even the gods would have trouble coping with.

In a famous—and baffling—statement, Einstein asserted:

> The distinction between the past, present and future is an illusion, even if a stubborn one.

Isn't that statement alone a strong indication that his idea of time had led his thinking into a surreal world?

Intuitive reasoning tends to transform puzzling features of the universe into "illusions" that fit better in equations and abstract speculations. In many cases, the illusions turn out to be metaphorical products of the mind that have little or no relationship to anything in the real world on the other side of that existential line.

For most of the 20th Century, relativistic and quantum illusions or "entities" were blended together in Einstein's unimaginable, four-dimensional spacetime continuum. In recent decades, however, theorists have been swamped with more "things" than can be fitted into only four dimensions, so they have devised supercontinuums defined by 10 or even up to 26 dimensions of space and *time*—realities they think actually exist outside their minds.

13

Human Reality

The meaning of time is the meaning of life.

To better understand who and what we are and why we can give meaning to our lives, I think we should look more closely at how we connect ourselves to perceptions of the exterior reality—from an objective standpoint supported by scientific studies and that good old common sense.

Nature evolved our sixth sense of time and all the other functions of the brain to establish an act-and-react relationship between the mind (the brain in operation) and the non-conscious material world where lots of things are happening.

In mythical terms, we could say that Nature gave us two stacks of materials to build our part of the relationship. In one pile—matter, energy and space. In the other—the senses and creative powers of the mind. Then she said:

> Okay, my humans. Here are all the tools and materials you will need. Go ahead and build your own world in the land of human consciousness.

The relationship is suspended when an individual's brain is completely unconscious, as in a dreamless sleep (or death), and then the motion-detection system ceases to operate for that person. In another machine analogy, the brain is like a motion picture projector that has been turned off. The basic components are still there—information recorded on film, a light source, a driving mechanism. But only when the machine is operating can it flash changing images from the screen.

Similarly, the brain "turns on" when its time system starts delivering sequences of perceptions and thoughts—mental strings of electrochemical energy flashing over a complex network of neurons where information has been recorded.

If we accept the premise that something cannot be created *ex nihilo* (out of nothing), we come again to our earlier conclusion that consciousness and time are produced by interactions between energy and neurons in the physical substance of the brain.

Or to put it in poetic terms, the energy of life caresses the silken receptors of our inner selves and gives us the feeling or awareness of conscious, thinking beings.

The Red Disk

Now for a look at key factors in how we create our worlds on the mind's side of the existential line. Say you're studying a large white wall on which hangs a small red disk. The molecular surfaces of the wall and disk reflect light into your eyes, and your mind adds on the colors—white (all of the wavelengths of "visible" light) and red (just the wavelength that stimulates the red color cones in your retinas).

You might think of a stream of colorless water entering your eyes, and your retinas injecting a white or red color as the "water" flows on into your brain.

(A friend told me the other day that cats don't have red cones in their eyes, and if that's so, I guess they see mixtures of the other two primary spectral colors. Can you imagine a world painted in only blue and green?

But how come the colors seem to be painted *on the wall and disk*—not on some screen in your mind?

We could say that what you see at any instant is a combination of incoming data and mental qualities, which gives you the impression that what you observe "now" is what exists out there "now." However, recent research indicates that it takes several milliseconds for the brain to absorb the incoming data, interpret and embellish it with past data, record it all in short-term memory cells, and then replay the scene in another complex of neurons—in Technicolor™.

So, from moment to moment, what you're watching are actually pictures projected on the inner screen of your mind, while other parts of your brain continue to take in more information.

> The world *you see* at any instant is totally a creation of the mind! Isn't that marvelous? Doesn't it give you a sense of your vital, important role in whatever exists and happens in the world around you and in the universe?

Seeds of the Soul

Other research suggests that consciousness is not continuous. Instead, it flickers on and off between segments of the mind's ongoing perceptions, memories and thoughts. Self-awareness clicks on when the mind focuses momentarily on one "now," and as the number and variety of "now points" rapidly change, the resulting impression of continuous consciousness is analogous to a movie viewer's perception of constant motion. Of course, the viewer is actually watching a fast sequence of momentarily fixed light patterns projected on the screen from film frames.

When in deep thought or daydreaming, the mind roves over recorded information and mingles memories with speculations about what the future holds in store. We exclude immediate awareness of the world around us, and the "present" in our minds becomes a shifting back and forth between the remembered past and the imagined future.

The brain is rather like a video camera equipped with both recording and projection systems and its own screen. But I would never say the brain is a machine. Never. Analogies can help us better understand how the brain works, but that doesn't make it a machine.

Human consciousness, which embraces everything the mind can perceive and think and feel, transcends anything we can ever learn about machines.

> From the evolutionary view of things, we can think of the brain as the garden in which Nature planted the seeds of the soul. We may not see many flowers blooming in the garden yet, but we can still hope the immortal soul may be the greatest of all the possibilities of life, on this earth and in the universe.

Adding Time to the Picture

Now suppose that red disk starts moving across the wall. Your brain records (memorizes) a sequence of disk positions relative to the sides of the wall, and the very process of handling that information and replaying it in your mind is what gives you the sixth sense of time passing—and your self-awareness.

We experience this connection between physical movement and sense of time in countless ways. Say you're out driving at night in the country on a smooth, paved road. The rear view mirror shows you the road behind you. A glance out the side window reveals the terrain where you are now. Looking through the windshield shows you the road stretching out ahead where you're going.

Your awareness of continuous physical movement comes from those three views—where you were, where you are now, where you're heading.

But what if it's a very dark night and your headlights suddenly fail? The road and terrain would completely disappear from your conscious awareness. Without reference points or markers, you wouldn't know anymore whether you're moving or not.

You would seem to be "frozen in space."

Similarly, your sixth sense of time comes from three segments along the "road of life"—your memories of what has already happened (the past), your awareness of what's happening now (the present), and your expectations of what may be happening next (the future). But if your memory and imagination were suddenly blanked out, you could no longer string the three segments together in a sequence.

You would seem to be "frozen in time."

The mind creates its sixth sense of time through a synergism of memory, present awareness, and thoughts of the future. By arranging remembered events before and imagined events after our immediate awareness, we create a larger continuity in which time seems to flow from the past into the present and on into the future.

In our minds, subjective time can skip around, we can move it backward and forward in our memories, rearrange the fragments in any number of different sequences. In relation to objective reality, however, time is a highly sophisticated, biological motion sensor in our brains that perceives, memorizes and measures the inexorable, ongoing chain of events driven by gravity and three other natural forces on the other side of the existential line.

> The mind's time and consciousness pervade this earth and the observable cosmos; there are no empirical traces of those wonders in a brainless material universe that exists in perpetual and eternal darkness.

14

Evolution of Time

Time is not a *thing* like St. Paul's Cathedral.

Of all our inborn abilities and instincts, none could be more fundamental than a neural system for keeping track of what's happening in our lives. The evolution of the brain would have to start with a basic motion-detection system, a framework on which the experience of being alive could be memorized and recalled in a coherent sequence of events. Otherwise, the mind could never have discovered any order and grace in the way that skeleton performs his dance.

Time as a mental function began in the brain stem, then evolved into more complex faculties in the reptilian, limbic and cerebral cortex areas. Research indicates that the highest form of time is in the cortex's frontal lobes, where we consciously anticipate future events and use our "working memory" to figure out what we're going to do next.

Rates of Time

Trees, fields of corn and other forms of non-conscious organic life obviously have no sense of time; if they did, they would be conscious. (For heavens sake, don't anybody tell that skeleton about time; if he finds out, he'll become conscious himself and start spying on us for a change.)

In the animal world, monkeys, dogs, birds and other forms of conscious life have brains and senses much like those of humans. However, their motion-detection system is governed largely by instincts, rather than thoughts, and it operates at a much simpler and less flexible pace. The "standard" rates appear to vary among the species. For example, a hummingbird lives at a fast-moving clip that's difficult for us to follow, whereas a turtle's time just pokes along.

Unlike a computer, humans don't have an absolute pulse rate for keeping time; if we did, we would be robots and we wouldn't need any clocks and calendars to correlate our activities and interests. Instead, our timekeeping system changes with our state of mind and the conditions in our lives. As we age, time seems to pass faster. When we fly to London or some other distant location, our mental clocks have to adjust to jet lag and a different time zone.

In ordinary usage, time and change (the result of motion) are interchangeable perceptions. Time is change. Change is time. Then we expand into the figurative: "Time has left its mark on her beautiful face."

Defects in Time

Defects or breakdowns in the brain's timekeeping system cause mental dysfunctions. Perhaps the most crippling effect, in otherwise normal people, occurs in cases that psychologists call "motion blindness." The afflicted person can see the world clearly when everything is perfectly still (frozen in time), but not when things are moving.

As a result, the individual perceives only a jerky sequence of images that produces an unpredictable and often frightening view of what's happening in the world outside the person's mind. The person dare not venture out without the guidance of a nurse or other caretaker.

> The rare disease clearly indicates the normal, fundamental purpose of the brain's timekeeping system in coordinating

mental perceptions with motion in the exterior reality outside the mind.

In schizophrenia, a person has extreme difficulty in distinguishing between mind and reality, a pathological problem that may lead to aggressive behavior because of the brain's damaged ability to connect the subjective to the objective. Autism appears to be a generally passive form of that inability to relate perceptions of the exterior reality with the ongoing sequence of thoughts in the mind, and the person may become obsessed with "stopping time" and its irritations.

Dyslexia, a reading disability that afflicted Einstein, may be a defect in the brain's ability to play back and interpret sequences of visual perceptions in the order they were memorized in various areas of the brain. Research indicates that attention-deficit disorder may be an imbalance or shortage of neurochemicals that the mind needs to focus selectively on particular "moments in time."

The brain's motion-detection system is also a major factor in an individual's intelligence and talents, which could be defined generally as the ability to memorize sensory information accurately, recall it coherently, and utilize it in ordinary or creative ways.

Time organizes and manages the symbiosis of subjective and objective realities in the mind, and as psychologists and neuroscientists learn more about the time function, they will better understand the evolution, basic structure and operation of the brain.

The knowledge could lead to new treatments for mental and emotional disorders.

Don't Forget the Skeleton

Besides effects from diseases, the brain can be damaged or thrown out of kilter by other things people experience in life—a knock on the head, emotional upheavals, too much booze or drugs. Then the pictures on the inner screen of the mind become blurry, confused, surreal fantasies. Even when the mind is at its clearest, ideas from the imagination can give us a distorted picture of what actually exists in the world on the other side of the existential line.

The idealist philosophers (and later the quantum theorists) were partly right. Subjective reality, which contains images of physical reality, certainly wouldn't exist without our minds. Therefore, the universe *they perceived* could exist only in their minds. However, they left out the separate material world, the skeleton of matter and energy that danced unnoticed for aeons in cosmic darkness and silence, until spectators evolved minds and senses capable of penetrating the darkness and started observing his performances. The old scholars didn't have a "scientific camera" that could keep track of all those trees falling in the forest when nobody was around.

The sophisticated cameras of today record pictures of the real world, but unfortunately the images can be blurred by that mixup between mind and reality. Theorists and researchers tend to project their own ideas and metaphors on the "big screen of reality" and pretty soon they see all sorts of wonders and horrors, such as "black holes," alien "dark matter" and whole universes exploding out of infinitesimal points of space and *time* (i.e., somebody's timekeeping brain functions).

Raise the Banner

In designing our minds and senses, Nature foresaw the human need to take a break "from time to time." So we can close our eyes, tune out our ears, and disconnect from that sometimes harsh reality out there. Then we can draw on recorded memories and watch our own "videos" shown on the inner screen of subjective reality. Daydreams...fantasizing about that sexy woman or handsome man...imagining various ways to handle a problem at the office.

But Nature inserted another function in our brains called ambition. "That's enough time-out, you guys. Get back to work in the real world."

Then, with our ideas and feelings refreshed, we are galvanized into action and we get busy changing what we can in the world.

But now let's pause here on the Quo Vadis trail, look off there into the far distance and catch a glimpse of what we can fancy as the grand purpose that Nature has given to intelligent life:

> Climb aboard your Vessels of Time, my humans, raise the banner of humanity, and set sail for the stars. Break the cosmic silence and bring light and meaning to my universe.

Does Anything Matter?

In discussing these *quo vadis* issues with people in the past, I would usually start by noting that time is a word we apply to motion and change. Nothing new in that. The definition goes back at least as far as Aristotle, who said, "Yes, time is motion," and then he went on to say that time is also an intrinsic, fundamental part of the physical universe (apparently, Aristotle's scholarly approach excluded any touch of our "childlike innocence").

The amused look in the eyes of my listeners (not you folks walking along the trail with me now) would turn blank or withdrawn when I went on to say, "Yes, but time does not exist outside our conscious thoughts."

Most eventually accepted the meaning of time as the mirror image of the conscious mind, not a substance or dimension "welded" to space, but they seemed uninterested in the next challenge:

What were the *consequences* for them and the world community?

The mighty, century-old tower of Einsteinian science was crumbling before their very eyes. Couldn't they see that flare in the night? Shouldn't bells be ringing and voices shouting?

> My God, Einstein and the quantum theorists must have been wrong after all, so now we can brush aside these degrading stories of who and what we are and seek a fresh, hopeful vision of life, on this earth and in the universe.

How will the work of quantum scientists be affected? Should they continue with their present and planned projects? Will any of them apply for that million-dollar reward? You mean we've got six senses, not just five? Have the biologists heard about it?

Well, actually, they seemed to be thinking, we don't have any expertise in science, so how can we make judgments on these complex matters? Besides, the world is changing so fast now it's hard to tell what's real anymore. Let's wait and see what the recognized authorities have to say about all this.

Meanwhile, if we get tired of hearing about universes exploding out of pinheads of space and *time*, we can rely on the science-fiction writers to tell us more believable stories of what's happening in the cosmos. And if time tunnels have been closed to further travel, the writers will find another way to get their characters to places where no one has gone before. Excuse me, I have to check in with my stockbroker before the market closes.

Pardon me. Your trail guide should know better than to nag at people. The volume in this world keeps turning up, people's minds are overwhelmed with daily floods of news and entertainment, appeals and demands. The last thing they need is another bearded prophet shouting at them on a street corner. Come to think of it, I don't even have a beard.

I'll pause here and listen to one of my favorite melodies. *Largo* from Handel's opera "Xerxes." Such a soothing, sweet sadness. I do care about life and its future.

Part Three

Age of Intuitive Physics

15

Baking Pies And Continuums

> The deepest human need in certain generations has not been to make sense of our human lives, but to make nonsense of them.
>
> Philosopher Archibald MacLeish

As we proceed down our Quo Vadis trail, we'll find more cultural, philosophical and spiritual reasons for rejecting the metaphysical messages of 20th Century science, but first I promised you a review of the evidence that seems to support current theories of the physical reality outside our minds.

We'll move briskly along this next stretch, just touching on the more significant concepts. "The rest are details," as Einstein once said in presenting his basic views. Besides, I'm eager to get your reaction to ideas of *space* that to me have always raised some of the most fascinating questions about the origin, nature and meaning of our existence.

There are some questions that historians, psychologists, sociologists and other scholars may want to explore more deeply than we can manage on this journey. For example:

Why did some of the most brilliant men and women of this era want to create their new realities and how did they impose them on Western cultures? And that intriguing—and troubling—related question: What was there about the past century that made people willing to accept passively a radically different system of ideas that told them their lives and all human existence were meaningless?

When cosmologists announced that the whole universe exploded out of a pinhead of space and *time*, why didn't the whole world erupt in laughter? Why didn't the universities call those geniuses back to school for a course in Reality 101?

Freezing the Universe

Advocates of Einstein's theories often cite his equation $E=mc^2$ as proof that his concepts of physical reality are truths of Nature; the equation was, they believe, the key that opened the door to nuclear energy. But let's use our Time Key to unlock another perspective on the relationship between Einstein's space-time concepts and technological advances in the 20th Century.

$E=mc^2$ is an idealized form of complex equations that express the *theory* that mass (m) converts into energy (E) and vice versa, with the amount determined by a proportionality factor—the square of the absolute speed of light (c^2).

As it turns out, the equation is by no means simple or even consistent. The *m* actually represents *two* masses: one when a planet or other body is in motion, and the second (m_0) when the planet is at "absolute rest"—an impossibility in Einstein's universe where *all* motion of physical bodies is supposed to be relative to each other.

Any quantity, say the number 10, has no meaning unless we can relate it to a marker or peg, like 0, that never changes. So Einstein had to freeze his universe mathematically long enough to calculate time and distance relative to some "zero" that wasn't moving. Early skeptics called the "transformation" sleight of hand.

Further, his assumption that the velocity of light is an absolute constant, with respect to all moving objects on which "observers" are perched, requires that time speed up, slow down or reverse itself in order to maintain light's speed as a constant in his equations. In effect, he replaced the classic absolute of immovable space with his absolute of light's velocity.

> When physicists and neuroscientists confirm that time is solely a complex biological motion sensor in the mind, and not a dimension or physical material "welded" to the spatial void, Einstein's concepts of the nature of light and how it is transmitted through space will have to be replaced by "timeless" ideas.

Coming and Going

In theoretical terms, $E=mc^2$ did fit into independent speculations by other physicists on the possibility that huge amounts of energy were locked in the core of atoms, and experiments were showing that such energy could be released through nuclear chain reactions. Einstein knew nothing about that empirical phenomenon until after laboratory tests indicated its feasibility, and his idealized equation is off the mark by a factor of nearly 100 in predicting *how much* energy would be released—at least in the nuclear reactions that have been devised so far.

In practical terms, it could be reasonably argued that Chadwick's discovery of the neutron and other work by empirical scientists—not Einstein's theories—produced the wonders and horrors of the Nuclear Age.

As for quantum theories based on Einstein's ideas of space and *time*, they do provide highly accurate mathematical models at the atomic and molecular levels, and formulas derived from them have yielded many practical results in areas of solid-state physics. Otherwise, the theories would have been discarded long ago.

The Ptolemaic earth-centered model also would have been thrown in the dustbin of history long before Copernicus, if it hadn't worked fine as a navigational tool and predictor of celestial events for about 14 centuries.

That's the way it is. Theories come and go, and I'll bet that, *sans* Einstein, engineers and empirical scientists would still have invented computer chips, lasers, cyberspace, robots and all the other marvels of the Age of Technology—and without many theorists spending their careers painting a celestial picture filled with weird black holes, dark matter, superstrings and "parallel universes of higher dimensions of space and *time*."

The Mercury Puzzle

Some of Einstein's other equations fit mathematically with several phenomena, such as the tiny "precession" in the planet Mercury's orbit. Newton's principles of motion and gravity accurately track the orbits of planets farther away from the sun, but founder a bit when applied to Mercury's path. Could there be another *natural* force that influences the planet's motion?

Some physicists have speculated that electromagnetism may play a small but significant role in the motion of stars and galaxies, so it may be that the close proximity of little Mercury to the huge sun produces electromagnetic force fields of sufficient strength to affect gravity's determination and control of the planet's orbit.

If Einstein had not brought the spacetime continuum into the picture, physicists and astronomers may well have followed the Newtonian path until they had devised theories and equations that solved the puzzle of Mercury's precession—and left us with a real and comprehensible universe.

Bending Light and History

Another phenomenon that seems to support Einstein's theories is starlight's very slight deviation from a straight path when it passes close to the sun and other massive bodies.

Everybody had long known that light bends when it passes through water, prisms and other substances, but it remained for Einstein to devise equations that predicted light would bend by about one-thousandth of a degree when it traveled close to the sun's powerful gravitational field.

The match between Einstein's equations and the light's angle could be attributed to his extraordinary talents and, perhaps, a historic stroke of personal luck. If his equations had failed to fit that miniscule bending, as later measured by astronomers, his visions of the reality outside the mind would never have risen to dominate scientific and cultural thinking in the 20th Century.

However, theories are not, by definition, established truths of the natural world, and in a sequel to this book, we'll review other concepts that could account for the sun's power to bend light rays—and once more leave us with a universe we can believe in and understand. For now, I'll bring in a quote from astronomer Carl Sagan:

> The history of science is that conventional and accepted ideas are often wrong, and that fundamental insights can arise from the most unexpected quarters.

Twilight or Dawn?

In order to move on in new directions, we must first insist on a basic fact of the relationship between mind and reality: Equations can never, ever, create a spacetime continuum or anything else. The equations are not the reality. The reality is not the equations. When we can no longer distinguish between the tuxedo and the real or imagined body it fits over, science and metaphysics spin off into fuzzy dream lands.

Cosmological models can only, at most, *symbolize* or *represent* certain features of a real universe. So do the abstract models developed in the 20th Century accurately represent the cosmos? Is the quantum paradigm a "magnificent framework" for comprehending the universe "and our place in it," as contended by science writer John Horgan in his 1996 book *The End of Science*?

Have the physics theorists been so "spectacularly successful" at describing the principal features of the universe that they have now reached the "limits of knowledge"? Or have the theorists merely bumped into the outer limits of what can be fabricated out of exotic equations and illogical, impossible notions of space and *time*?

> Do we stand before Horgan's melancholy "twilight of the scientific age"? Or will we turn to objective, comprehensible theories of a real world and see the dawn of a new era of discovery?

The era will begin when we return to a reality in which 2+2=4.

No Dancing Skeleton

"Every new law [of physics] must be tested against Einstein's theory," says physicist Kip Thorne in his 1994 book *Black Holes & Time Warps: Einstein's Outrageous Legacy.* Let's run down a list of "laws" again and see if there's anything left of that "outrageous legacy":

> Since time is not a "substance" that can be mixed and compressed with space, there are no miniscule "black holes" out there, at least not in a "fabric" made of space and *time* (i.e., somebody's brain functions).

> No universes exploding out of infinitely small particles and creating their own space and brain functions.

> No relativistic speeding up or slowing down of brain functions.

> No machines to carry time travelers back and forth in a stream of brain functions, except in science-fiction tales.
>
> No planets, stars and spaceships following "warps" in a continuum composed of space and brain functions.
>
> No cosmological constants woven into the fabric of space and brain functions.
>
> No gravity transmitted by "ripples" in Einstein's space-brain functions continuum.
>
> No cosmic superstrings or membranes vibrating in higher-dimensional universes of space and brain functions.

Since time is not a structural "dimension" that can be bonded to the dimensions of space, cosmologists have no bricks and mortar for building their antiworlds on the other side of the cosmic mirror; no beams and girders for constructing those invisible, empirically undetectable parallel universes that they say created and now supply this world with its mass, gravity, light and a growing number of "virtual" particles.

Suppose that all of a contractor's conventional tools and materials for building homes suddenly disappear, vanish, they no longer exist. He would be out of business overnight, wouldn't he? Or to put it another way:

> If pumpkins did not exist, you couldn't bake a pumpkin pie. If time does not exist as a substance or dimension outside our minds, you can't cook a space-time continuum. If the space-time continuum does not exist, there is no framework of physical reality, no dancing skeleton on which to hang the flesh and blood, and the heart and soul, of Einstein's universe.

16

Truth And Beauty

> Let us worry about beauty first and truth will take care of itself! Such is the rallying cry of fundamental physicists.... In the silence of the night, they listen for voices telling them about yet-undreamed-of symmetries.
>
> A. Zee, *Fearful Symmetry*

Some cosmologists and quantum physicists (but certainly not all) have completely abandoned the classical approach to discovering the truths of Nature; that is, start with empirical evidence of what seems to be a feature of physical reality, and work from there to draft theories and equations that may help researchers better understand the phenomenon.

Instead, the "fundamental physicists" rally around an abstract mathematical model that expresses their ideals of Nature's "beauty and symmetry," and then, like the medieval astronomers working with the Ptolemaic design for the heavens, they spend their careers endlessly adjusting the model to fit any empirical evidence that may support or contradict it.

"Beauty is truth, truth beauty," poet John Keats concluded. "That is all ye know on earth, and all ye need to know."

Many would agree with the Keatsean principle, when it is applied in the world of art. But if science is to speak to us in poetic languages, surely the cosmologists can come up with a more beautiful description of their universe.

Do you see any beauty in a cosmos that begins with a Big Bang and ends in a Big Crunch—especially when those faraway pictures are shrouded in perpetual and eternal darkness?

The Black Hole

Decades of costly experiments have failed to yield any direct evidence that the spacetime continuum actually exists in physical reality, but the effort continues to this day. In one research project, "gravity observatories" are being constructed to detect "gravitational ripples in the fabric of space and *time*."

Cosmologists, in explaining the purpose of the quarter-billion-dollar project, don't talk about the continuum as a theoretical model. For them, the continuum is a vast medium, a celestial ocean that fills the universe; it is the "where" in which they think everything exists and swims around in obedience to their equations.

In another research project, a satellite will be launched into orbit to detect how the spinning earth drags the spacetime continuum around with it, like a net. The "net" is not presented in news reports and science journals as an abstract model or metaphor of gravitational interactions between our planet and the rest of the universe. For cosmologists, the curved lines depicted in their drawings actually exist in the "fishnet of space and *time*."

Misconceptions of time and space have locked quantum cosmology and particle physics into theories that cannot be applied productively in these and many other projects. Yes, the universe is very mysterious and complex, but playing around with abstract models and poetic fantasies doesn't bring us any closer to "the truth."

> We are only pushed deeper into a black hole of absurdities—the Omega Point for human thought that can no longer distinguish between mind and reality, the final end for intelligent beings when the poison of meaninglessness seeps into the blood of life.

The Particle Quest

The Superconducting Super Collider project in Texas was to have been the climax of decades of searching for phantom particles that flash in and out of "higher-dimensional" worlds in a few trillionths of a second. However, faith and funding waned after four years of construction, and Congress canceled the project in 1993.

If the project had been completed at an estimated cost of $11 billion, it would have led to "a major discovery, as major as any in this century," according to Princeton physicist David J. Gross. "It would have enlarged our view of space and time. It would have proved the existence of other dimensions."

With all due respect to Professor Gross, a rational analysis of the symbiosis between mind and reality leads to a quite different "major discovery":

> Einstein's time and "other dimensions" exist only in the thoughts and equations of the theorists. Consequently, all theories and research based on false concepts of time and space are a waste of public money and the talents of many brilliant men and women.

In human life on this earth, ideas can lead to new business and social enterprises, improve or devastate our cultures, or bring about fundamental changes in the planet's natural environment. The cosmos, however, is beyond the reach of human thought and activities. It is what it is, and it is totally indifferent to what we think about it.

But when we take out our existential pens and draw a line between mind and reality, we can more clearly differentiate between our ideas and what actually exists. Ideas help us identify and understand what's out there, but they cannot create or control universes.

The ideas are not the universe. The universe is not the ideas.

Just as time, a creation of the mind, is not motion.

Motion, a feature of physical reality, is not time.

"Scientific Odyssey"

Where did these ideas come from? In his 1994 book, *Hyperspace: A Scientific Odyssey Through Parallel Universes, Time Warps, and the Tenth Dimension,* physicist Michio Kaku seems to acknowledge that faith was the source of scientific as well as spiritual beliefs in the 20th Century. In his own words:

> [Some] people have accused scientists of creating a new theology based on mathematics; that is, we have rejected the mythology of religion, only to embrace an even stranger religion based on curved space-time, particle symmetries, and cosmic expansions. While priests may chant incantations in Latin that hardly anyone understands, physicists chant arcane superstring equations that even fewer understand.

According to Kaku, "The 'faith' in an all-powerful God is now replaced by 'faith' in quantum theory and general relativity."

If that is so, then we could point out that faith *in any form* calls upon the same emotional needs and commitments to sustain it. The more devout physicists cling to beliefs as fiercely as any priest or minister, despite the dearth of empirical evidence, and they use strikingly similar rituals to scare off skeptics and gain the allegiance of people who can't understand their "arcane" incantations.

Theorists don't want to hear that there is no spacetime continuum, any more than the medieval monks wanted to hear that there is no God.

Without the continuum, they would have to end their costly pilgrimages to make-believe worlds and return to the land of reality.

The mission of science should be to explore the many remaining mysteries of the universe and discover the treasures still buried in a reality we can all believe in and understand.

> Realistic science can also help us accept the terms and conditions laid out by Nature for the continued survival of the human species.

Period of Upheaval

Meanwhile, we should reject these assertions that our poor human brains simply have not evolved enough to comprehend bizarre quantum theories. From a skeptical viewpoint, such claims are a way of escaping accountability for pouring talents and money into exotic projects that supposedly transcend ordinary standards for judging their worth. In his book, Kaku reports:

> [The theory] of 10-dimensional hyperspace has already swept across the major physics research laboratories of the world and has irrevocably altered the scientific landscape of modern physics, generating a staggering number of research papers.

Don't you, as a taxpayer and private citizen, also feel a bit "staggered" by such expenditures of money and talent?

If we could snatch time out of the mind, the result would be catastrophic; any kind of order in life would become impossible. But when we snip time out of Einstein's space-*time* continuum and stick it back in the brain where it belongs, the universe won't collapse in rubble. Matter and energy, governed by Nature's own "timeless" laws, will just keep on rolling along.

There will be a period of upheaval as current models are discarded and new theories and equations are devised to fit a rational, comprehensible universe in which matter, energy and space are the only components on the objective side of the existential line that divides conceptual possibilities from what actually exists in physical reality.

Now I get to tell you about some metaphysical aspects of two other factors that have shaped scientific and cultural views of life and the universe: space and dimensions.

17

What Is Space?

Why doesn't God or Nature tell us the truth, the whole truth and nothing but the truth?

In one of his papers, Einstein came to a "logically unavoidable" conclusion that he said had eluded "pre-scientific thought":

Space is broken into an infinite number of separate, distinct areas or "boxes," which are in motion with respect to each other.

Quantum mechanics then went on to divide up space into discrete, infinitesimal "loops," or in other theories, space was laid out in "layers" that could be rolled up like carpets. What exists *between* all these separate pieces of space? Apparently, the pieces have to float around in the quantum "vacuum," the latest medium for motion.

Let's go back to the classic concept of space as an absolute void, a "pre-scientific" view that prevailed until the 17th Century, when Descartes started the long trend of replacing space with things like his "luminiferous aether." Greek philosopher Leucippus defined space as the "where" in which everything exists and moves around, but now to simplify things further, let's divide the cosmos into two universes—the material and the spatial.

The material universe, composed of matter and energy, occupies just portions of its spatial "home." We might think of it as a transient who keeps moving around and never has to pay any rent to his "landlord." The material universe could be finite since we can imagine boundaries in anything tangible and real, and other finite universes may exist over the cosmic horizon, near or far.

In a conceptual sense, each universe must be eternal because, contrary to the claims of theorists, we can't have some real thing emerging from, or vanishing into, absolute nothing (except in ghost stories).

What Is Where?

As an absolute void, the spatial part of existence could have no physical properties that would affect the motion of planets, stars and other forms of matter and energy. Neither could the planets and stars have any effect on the space in which they exist and travel around.

> Space is immutable, the same everywhere, a dimensionless void that can't move, shrink or expand, and it could have no imaginable boundaries. Space can be measured only with respect to reference points in the material universe.

But where does that definition lead us? To another, even more profound "something that thought cannot think"? Sure, space has an obvious purpose as the "where" in which everything exists and moves around, but *what* is space? What about sectors of the vast universe where no forms of matter and energy exist and move around? Does the "where" exist in those places too?

We could define space as the ultimate nothingness, yet "something" must be *there* to give us our *where*.

But *what* is a thing that can be defined only in terms of its purpose? Can you think of anything else whose objective existence can be known through its purpose—but happens to be totally invisible to our eyes and instruments, and hasn't a single detectable feature or property?

A Spacey Feeling

The idea of an absolute spatial void teases the imagination but finally confronts us with what may truly be the limits of the human mind. We exist in the spatial universe and therefore space itself must exist—right? But could space exist without us and that material world on the other side of the existential line?

Well, it's not a co-dependent relationship. We need space as the place where Nature or God assembled "all the choir of heaven and furniture of earth," but space doesn't seem to need us; it doesn't vanish when it's not occupied. Your home and the space it encloses don't depend on your presence for their continued existence (except for upkeep and paying the mortgage).

To venture deeper into the depths of emptiness, let's try out one of those "thought experiments" that theorists like to play with. We construct our own "space box," one foot square on each side, and give it some special features: The walls of the box ward off gravitational forces, cosmic rays, light and all other forms of electromagnetism, and we'll suck out even the tiniest particles of matter and energy hiding in the box.

Voilà, we have a box that contains one cubic foot of the absolute void of space.

Now let's imagine that our box is also capable of expanding indefinitely, and as it does so, the walls push away all forms of matter and energy. The box expands to the size of our solar system, then our galaxy, then a hundred galaxies, then a million, a billion, a trillion.... Let's go for broke and enlarge our box into "infinite" size.

But infinity, by definition, is a size bigger than any we can conceive. Can you imagine a box that contains all of the space that our minds are capable of conceiving, and on the outside of the walls, there's...what? Similarly, we bump into the limits of the mind when we wonder:

> How and when did space come to be, and what existed before there was any space? Did space nonexist before it existed? But

> *what* is nonexistence? Or has space been around for all of eternity? But *what* is eternity?

Hiding the Truth

In his speculations on the incomprehensible, medieval theologian and philosopher St. Thomas Aquinas also confronted the age-old *quo vadis* questions of where we came from and where we're going. He began by noting that humans can't imagine a beginning for life that has nothing before it, or imagine an ending for life that has nothing after it.

In other words, when you draw a line of any length, there must be something before the beginning point and something after the ending point (unless your line soars off into mathematical warps and ripples in the spacetime continuum).

Therefore, Thomas concluded, God must exist because only a supernatural being could create "something" out of "nothing," which has no beginning, and make it last for "eternity," which has no ending.

But what existed *before* God in the chain of existence? Nothing? Nonexistence?

If his God was not somehow the "First Cause" of all existence, then Thomas was merely cloaking the unknowable in the incomprehensible—a "solution" to mysteries that physics theorists discover when they try to venture beyond the boundaries of the intellect.

Let's ask the question again but from another angle:

> How come some parts of existence are totally unimaginable and unknowable? Why don't we exist in a universe where everything, from here to eternity, would be completely comprehensible to our minds?

Would we be better off in a universe that eventually told us everything about itself and answered all our questions? Or would life become boring and pointless if all the mysteries and uncertainties were eliminated? Perhaps humans will always need the unknowable to inspire the faith that no matter what outer boundaries they reach, there will always be one more frontier to explore.

A longing for visions of faith seems to be inherent in human nature, and sometimes, such as in periods of grief or fear, it becomes much stronger than the opposite desire to acquire only objective knowledge.

Point of View

If there are any answers to such questions, they may be far off in the future. So for now I'll just bring up the *what* that imposes boundaries on even the most profound thoughts in the human mind. To think about anything we must first establish what writers call a point of view (POV). We must place ourselves, in reality or imagination, in the context of our story and what's happening. Without a POV, there would be no storyteller to tell the story.

Now let's see how far our POV can take us before it bumps into the mind's boundaries. Can you imagine a universe in which absolutely nothing exists? Eliminate trees, houses, mountains, stars, galaxies—everything. What's left? Nothing? No, because you, the observer with a POV, are still there trying to conceive a universe in which nothing exists.

> Try it sometime; if you can extract the POVs of all observers from your vision, you should send up a flare, for you will have accomplished something that has not been possible since the beginning of intelligent life.

Can you imagine your own death? If you're an atheist, your POV can make you a temporary witness to what you imagine will happen after you die—the funeral, family and friends gathered around your grave, former co-workers returning to their jobs, life going on as you sink into oblivion—the ultimate state of the forgotten that will last far longer than your point of view as a mortal observer.

If you believe in an afterlife, you can project your POV beyond the finite boundaries of this life and imagine yourself in the company of God and His angels for all eternity—an endless sequence of witnessed events (time) that will outlast your ability to maintain your mortal point of view.

If we say the hereafter is "timeless," we're merely substituting one unimaginable for another.

People report near-death experiences, but all of their images, such as a glowing light at the end of a long tunnel, are versions of what we see in this life; they offer no pictures or stories that extend beyond the reach of the mortal mind. Death is a fact of life, and with a POV we can observe or imagine a dead body's molecular structure disintegrating and finally vanishing from our sight into the earth.

> But what happened to the consciousness, the human spirit that once inhabited that body? It wasn't made of molecules and atoms, so did it vanish from our sight into...what? Heaven? Hell? Oblivion?

Only Miltonic poets and spiritualists can take tours of such realms and come back and tell us "all ye need to know."

We have our points of view now, and we can think about our births and place in the history of life. Our POV becomes the "eye of the camera" that shows us what's happening during the period of our own lives. When we die, our POV dies with us and the eye closes...forever? But *what* is a forever that we can comprehend?

> Nothing, inside or outside this life, can be perceived or imagined without a point of view. Even the most abstract thoughts require the thinker to place himself as a witness to whatever he observes, and when he ceases to be a witness, his thinking ends.

Trapped in Opposites

At the outer limits of the mind and its POV, we try to imagine things that seem to have no comprehensible opposites. Yet the universe reveals itself to us only within the boundaries of opposites that we can perceive and understand with our mortal minds.

Consider how the mind thinks in terms of counterparts. To perceive or imagine an "up," our minds must be aware of a "down." What would "hot" mean to us if there were no "cold"? We couldn't think of something as "heavy" if there were no comparison with something else that's "light."

Heaven is a place where everything is "good." But what would that term mean to angels if they were not aware of another world where the opposite of good existed? Does it mean anything to say that goodness is good and evil is evil?

Space has no conceivable opposite that could define it, so Einstein inserted a "time opposite" in equations and metaphors that ignored the enigmas raised by Saint Thomas. He wrapped the universe in a giant ball of space and time, and dismissed as "meaningless" any questions of what exists outside the ball.

Don't we all tend to get rid of vexing problems by dismissing them as unimportant or meaningless?

Most of what we experience in life does have opposites, although they hide from us sometimes. For example, the despairing existentialists thought of human existence as meaningless and death as nothingness. But they were writing about a "waste land" that had been stripped of everything that nourishes the human spirit. Their realities existed as the devastated opposites of an imaginable world in which life would have a vital meaning and purpose.

Life is the opposite of death, just as light is the opposite of the darkness on the other side of the existential line. Only our gods have a POV that shows them what awaits us after our mortal existence ends.

Then they tell us about it in terms we can relate to what we know about this life.

If there were no gods, we would invent them as our only means of trying to satisfy the human craving to know the unknowable.

Mystery of Mortality

Suppose your spouse falls into a deep, dreamless sleep while you are still awake. You know the brain cells in that dear head are still alive, busy regulating necessary functions of the body. But the motion-detection time functions of that other brain are inactive, inert; the sleeper is "frozen in time," completely without a POV that could see what, if anything, exists in the land of the unconscious.

From your point of view, you could define the sleeper's temporarily unconscious state as the opposite of your consciousness. But the other person, when awakened from a dreamless sleep, will never be able to answer your question: *What* is the counterpart of consciousness?

You can turn off the night lamp and observe that the light has vanished, it no longer exists. Your POV tells you that darkness is the opposite of light, which is one form or manifestation of the electromagnetic energy that exists throughout the universe.

Similarly, when the brain is "turned on," consciousness becomes a manifestation of electrochemical energy flickering over complex configurations of atoms and molecules. But when the "light" of the brain is turned off permanently and replaced by the dark of death, *what* is life's opposite in the physical reality outside the dead brain?

The permanent darkness of the universe?

Nothingness?

God's "fortress against the dark's insidious despair"?

> Life observes death, but death has no point of view from which it could observe life—or itself. All we really know about death is that it just keeps stamping The End on our POVs.

> Mortality is the ultimate mystery of life, the "supreme paradox of all thought." We come into existence and embrace the reality of this world, and then, after our brief turn is over, we vanish into... *what*?

The rational mind looks into the mirror of life and sees nothing more than reflections of itself. Only eyes blessed with the power of faith can catch glimpses of any worlds that may exist on the other side of the mirror.

Circling Around the Unimaginable

According to our Quo Vadis solution to the nature of time, all that exists in the material universe outside the human mind are matter and energy and a place (space) to put everything. That's objective reality, and our minds and senses make it real for us by dressing it up with light and color, sounds, taste, touch, smell, and time—the basic qualities of conscious existence that Nature gives us so we can interact with the exterior world.

We feel at home in the material world that our minds and senses tell us is out there, but we shrink instinctively from the black void of space, which has no features or properties that our minds can perceive and comprehend. So we imagine filling up the void with a "substance" that curves and warps in obedience to our equations, and we swaddle the observable cosmos with antiworlds and parallel universes that protect us from the chill of the cosmic night. Even our quantum "vacuums" are cluttered with all sorts of "particles" and "sparticles."

Such inventions of the mind occasionally produce mathematical constructs that can expand abstract speculations, but ultimately leave us stranded in the unimaginable and unknowable. We can imagine a universe that curves into a giant ball, but then we are faced again with that unanswerable question:

What's *outside* the ball? Infinity? Eternity? Nothingness?

But what do those words mean? How can we imagine "something" that has no boundaries, no opposite, no beginning or end?

> The only resolution, it seems to me, is to recognize that some aspects of life and the universe are indeed beyond human comprehension, no matter how many lines we draw with our existential pens. Then we can humbly accept the reality of space as one of the unsolvable mysteries and get on with the search for answers to the many remaining mysteries of the knowable universe. Space exists; we use it to frame any theory or awareness of physical reality.

When we try to go beyond the mind's limits with the help of ever more exotic mathematics and intuitive reasoning, we end up with a mish-mash of the only reality we can ever believe in and understand.

Hello, Alice

Physicists who contend that space, in the absence of all matter and energy, still contains something should accept the challenge of proving it. If their story is to be presented mathematically—without a clear distinction between mind and reality—we may suspect that they're taking another turn at creating "something" out of "nothing," an enterprise best left to the gods.

Equations do not a reality make. In fact, the language of mathematics can spin out all sorts of fantasies just as easily as fairy tales can be told in other languages used by humans. With adequate funding from a federal agency or institution that supports fundamental research, I could prove mathematically that Lewis Carroll's Wonderland really does exist—if not in this pro-world, then in one of those antiworlds or higher-dimensional universes composed of space and *time* (i.e., my mental motion sensor).

Hey, if cosmologists can shrink 200 billion galaxies into an "infinitesimal" point, I should have no problem adjusting the little girl's dimensions to fit rabbit holes and the tiny home of the March Hare. And when I get around to making the Cheshire cat disappear and leave only its smile behind, I'll use those equations that transport our world's mass into a primordial force field of 10 or more "other dimensions," leaving behind only infinitely small points that fit nicely in quantum computations.

> Any of the observable features of the universe can be defined mathematically. But when we invest numbers and symbols with the power to leap out from our equations and seize control of the universe and life, we abandon the home that Nature or God provided for us and give up any claim to a meaning in our existence.

Exotic equations and computer simulations of the cosmos will never satisfy the human longing for a higher meaning and purpose in life.

18

WHAT ARE DIMENSIONS?

> The paradox is only a conflict between reality and your feeling of what reality should be.
>
> Physicist Richard Feymann

The trail to Einstein's universe was opened by German mathematician Georg Riemann in the 1850s. Despite frail physical health and recurring bouts of mental illness, he devised a mathematical technique that enabled scientists to "think and theorize" in four dimensions, and thus break free of the conceptual restraints imposed for 22 centuries by two- and three-dimensional Euclidean geometry.

We can't visualize any world in other than three dimensions; our minds just can't imagine a fourth dimension sticking out at right angles to the three we use in defining the only reality we can ever know and experience (except in science-fiction tales). Mathematics works well within that limit and it has achieved many wonders over the generations, such as designing three-dimensional pyramids, cathedrals, Brooklyn bridges, shopping malls, high-rise office buildings, airplanes, space ships.

However, as Riemann discovered, mathematics can soar into higher dimensions, and a half-century later Einstein and his followers climbed aboard their own versions of the Riemannian equations and zoomed off into the mind-boggling wonders we hear so much about today.

Actually, when the theorists plug their equations into three-dimensional reality, the numbers and symbols erupt in gobbledygook. So they weld a few more "other dimensions" onto their equations, and suddenly a fresh vision of the beauty and symmetry of Nature rises before their eyes.

> Obsession with symmetry is the driving force behind the metaphysical side of quantum physics, and the theorists will clamber over any barriers of logic and common sense in their efforts to reach and embrace it.

Theorists have long felt comfortable in two-, five- and higher-dimensional worlds, and now another mathematical system has been devised that simplifies calculations in four dimensions. What about a little more clarity in human thinking and theorizing? Sorry, in that dull domain, physicists, as well as us ordinary folks still trapped in our 3-D brains, must keep muddling along as best we can.

> Sir Peter Medawar once defined science as "the art of the possible." In the 20th Century, "pure science" became the art of the impossible, a thinking technique that wiped out human reality.

"I am afraid of this word Reality."
Sir Arthur Eddington

Squaring the Hypotenuse

Georg Riemann was a sincere, courageous and very bright young man, and he came up with some interesting speculations at the outer fringes of the imagination. If the science and cultures of the 19th Century had felt a stronger need for such ideas, he perhaps could have become the Einstein of his era. Essentially, he redefined the meaning of dimensions, a conceptual leap into the 20th Century when theoretical physics would eventually eliminate virtually every distinction between mind and reality.

What are *dimensions*? Well, in practical terms, they are the framework of a system we humans invented to indicate direction and distance and size. One dimension (length) defines the distance between two points. Add a second dimension (width) and we have an imaginary plane that we can impose on a flat surface, such as a table top, to figure out its size.

If we rotate the plane 90 degrees, it returns to the concept of a one-dimensional line. To measure volume, we add a third dimension (height) and enter our real 3-D world.

In that Euclidean world, the three dimensions are arranged in a formal structure, starting with the idea of placing them at right angles to each other—like the straight lines at the corner of a box. The right angles are a handy way of designating three unique vectors from which all other possible angles can be measured.

So what we have basically are three "dimension-sticks" that we can arrange in various configurations to match what we observe about the size and shape of objects and the spatial distances that separate them.

Euclid, Archimedes and others figured out mathematical relationships among the dimensions. You recall that stuff from high school geometry (sines, cosines, tangents, etc.) that enabled you to calculate one side or angle if you could first determine other factors in how the "sticks" were joined together. Remember the one about "the square of the hypotenuse of a right triangle equals the sum of the squares of the two sides"?

Frames of Reference

Now we bring in brain time, a cerebral motion sensor that observes, predicts and calculates movement and change. Without mind-time, the 3-dimensional *model* could be applied only to completely static or freeze-frame situations, and in a universe of relative motion, those conditions can exist only in what's called a single "frame of reference." Nothing is moving in that lovely picture on your wall, so it is "timeless" within its single frame.

If we now bring in some handy observers with their own mind-time, things could be moving around in a single frame of reference, and with suitable instruments, one guy in that frame could use Euclidean geometry to calculate, for example, the velocity and trajectory of a baseball slammed by a batter. But another observer flying overhead in an airplane would come up with different figures because he's measuring the ball's flight relative to motion in his different frame of reference.

The point, of course: In the real world we know and experience through our senses, the timekeeping systems in the brains and instruments of the observers haven't the slightest effect on where and how fast the baseball is going (even if it's traveling at near the speed of light, relative to another observer).

Sticking Pins

So what's the only valid meaning and purpose of the "dimensions" of time and space? They are part of a system we humans use, in our minds and in the geometry we invented, to predict and measure events in physical reality and in our inner lives.

> Dimensions have no separate, independent existence apart from their purpose and use as practical tools by scientists and us ordinary folks. None at all.

Can you imagine a universe with 26 dimensions? No, of course not. Neither can the theorists. It's all "counterintuitive," which is the *opposite* (whatever that is) of the theorist's intuition. We can stick 26 long pins in a pincushion, and if the pins are bendable, we could curl them around, back into the cushion, and have sort of a round ball. But can you imagine more than three of those 26 pins, straight or bent, sticking in there at right angles to each other?

Everything from houses to universes is built of real materials; nothing in the physical reality outside our minds can be constructed of dimensional "toothpicks." The lines and curves and numbers in the drawings of architects and cosmologists are not what's holding it all together.

When we think of dimensions as a physical part of the material world outside our minds, we depart from reality and enter a fantasy land where clocks and yardsticks have the power to operate universes.

> And exploring nonexistent antiworlds and parallel universes of higher dimensions is a very costly venture, in terms of resources and talents and, more importantly, it deprives people of a worldview that matches what their reason and senses tell them is the real world.

No Dancing Skeletons

Riemann had his own ideas about dimensions. He thought they must somehow be built into the structure of matter and space, and so to provide more room for his visions, he added a fourth *spatial* dimension to the other three in his equations. It remained for Einstein (and novelist H.G. Wells) to extract the timekeeping system from the brain and insert it in the cosmos as a fourth dimension.

Social and intellectual elites in the 19th Century were fascinated by Riemann's idea of another world with four spatial dimensions, and they speculated about the creatures that might exist there (of course, without a "time dimension," those creatures couldn't be seen dancing around like our skeleton does). Several leading scientists even used Riemannian math to support the claims of psychics.

In the 20th Century, quantum theorists discovered they needed more than Riemann's four dimensions to define the new worlds emerging on the other side of the cosmic mirror. According to one theorist:

> The laws of nature become simpler and more elegant when expressed in higher dimensions.

Personally, I prefer the simplicity and elegance of the three-dimensional universe that Nature constructed for us long before the theorists arrived on the scene.

Meet the Flatlanders

Matrix mechanics based on the Riemannian model is one way of expressing the theorists' "laws." Mathematical points are arranged in a checkerboard of rows and columns, with each point corresponding to a place in space, or in the spacetime continuum, or some other "curved" realm that the theorists want to explore. General rules are established to govern how the points "bump and grind" as a sequence of numbers is fed into them, and "sooner or later" a pattern emerges.

Skeptics think of tensor calculus as a kind of Ouija board the theorists use to communicate with the higher-dimensional spirits who give them their visions of other worlds. The theorists, however, say the system conforms with the principles of mathematics, and therefore it must be presenting them with something that actually exists, even though nobody can "see" it.

To illustrate the supposed power of dimensions, the theorist may start with a two-dimensional plane where the "Flatlanders" are thought to live. We're invited to observe the activities of those creatures from our superior, three-dimensional perspective and try not to laugh at their confusion when the theorist wrinkles up their Flatland to confront them with the next higher dimension.

Similarly, we're told, more advanced beings may exist in four dimensions, and no doubt they are just as amused when we try to relate our 3-D perceptions with how things work in their world. But the theorist doesn't stop with four dimension-sticks. He inserts more and more sticks to build a higher and higher tower, hoping to see at each level an ever more fascinating image of the beauty, simplicity and symmetry of *his* Nature.

Magical Geometry

Now for one of those quantum leaps in logic that launched the "Age of Intuitive Physics." Since dimensions can now build and define whatever "universe" we're talking about, that must mean, Riemann and his successors deduced, that *geometry*—not forces of Nature—determines and controls everything that happens in physical reality, at least on the atomic and celestial levels.

Not figuratively. Not in the sense of an abstract model that can be used to represent observed phenomena. Einstein's general theory on the nature of gravity is all about geometry; it says nothing about natural forces. Here's how physicist John Wheeler once put it:

> Gravity is not a foreign and physical force. It's a manifestation of the geometry of space right where it is.

That kind of geometry has magical powers. Say, for example, the theorist starts by transforming the "up" dimension of an apple to the "down" direction. Easy: just rotate the apple 180 degrees. But then, the theorist thinks, time and space are also "dimensions," so that means I can "rotate" them in my equations and thus transform time into space, and vice versa, which means they're different forms of the same thing.

And since geometry can bend lines and surfaces into most any shape, that must mean that space itself is...bendable?

> Geometry is reality. Reality is geometry. Ergo, the universe exploded out of a pinhead of space and *time.*

Something=Nothing?

Nothing real can exist in two dimensions; there's no "where," no room for Flatlanders or anything else. So the theorist must use a little slight of hand here—he sneaks in an infinitesimal third dimension of height to provide some head room.

What's "infinitesimal"? Well, it would have to be a dimension that's unimaginably small but still big enough to create some of the 3-D space in which real things exist, but not so big that the Flatlanders would poke their heads through their ceiling.

How about a one-dimensional universe stretched out in an infinitely long line—or a world with no dimensions? Surely, you may think, the theorists can't find room for even Nonlanders in a mathematical point with no size, no space at all, not even any brain functions.

Actually, cosmologists routinely pack trillions and trillions of tons of matter and even entire universes into infinitesimal points, which are "something" defined mathematically as "nothing."

Or is it nothing defined as something?

The equations work both ways but always with the same result when their metaphysical side is applied to the human condition:

Humans equal out as nothing.

Amazing Grace

The world accepted Einstein's vision of reality—and the philosophical and cultural messages that it eventually implanted in the human mind—after the scientific community had incorporated his theories in a formal structure of experimental work and mathematical analysis.

Physics pioneers in the 21st Century could build a new structure by recognizing that the brain's motion-detection time system has no effect on the cosmos, and then turning their scrutiny and experiments back to forces of Nature that determine and control what's happening in physical reality.

When the world community is ready to say farewell, Dr. Einstein, people can start building a new reality of hope and meaning, a place where they and their descendants will be united in a quest to achieve a magnificent destiny.

Einstein's time and "other dimensions" do not exist outside the thoughts and equations of the theorists. Space is an absolute void.

So what is the basic framework of the reality outside our minds? It would seem that physics will have to establish a new framework for theories and research in the 21st Century.

What scientists discover in a new framework of mind and reality could help bring us all together in a song that everyone can hear and join in singing.

Words and music like those that still inspire the human passion for life. John Newton's "Amazing Grace"...Handel's "Hallelujah Chorus"... Beethoven's "Ode to Joy"...Puccini's "One Fine Day."

I eagerly await the day when composers and poets, inspired by their visions of *quo vadis*, start giving us more Songs of Life.

Part Four

Time to Rebel

19

Brave New Worlds

> Today every important new development in science causes the ground to shift, making ordinary people want to hold onto to something solid, if there is any such thing left.
>
> Cullen Murphy, *Atlantic Monthly*

Cosmologists and quantum theorists generally acknowledge that their phantom worlds of four or more dimensions can't be conceived by the mind, and some admit they really don't know, or even care anymore, whether their universes have any relation to physical reality.

So they use "projective geometry" and supercomputer simulations to transform the mind's images of a real three-dimensional world into "higher-dimensional virtual realities," and they invite nonscientists to join them in their enchanting realms of the imagination.

"Higher realities," one researcher says, "are what shape and define our complex cultures."

He may well be pointing the way the human race is heading: rejection of the real world and escape into the many virtual realities, from cyberspace to Hollywood movies, that science and technology have bestowed on mankind.

If you're not ready to join hands with your family and friends and stroll off into those Brave New Worlds, perhaps you may find some support in the doomsday satire of commentator Lewis Grossberger:

> After all, on our long, tedious march of regress have we earthlings really accomplished much besides spewing garbage, ammunition and the *Jerry Springer Show* into the environment? Wouldn't it set a nice example if we could just for once accept the inevitable and issue a press release reading, "Hey, we gave it our best shot, but we really weren't up to this existence business, so we're out of here"?

Time to Rejoice?

"Many scientists are saying we can integrate science into an existing religion, a personal philosophy of life, or New Age beliefs," says a Chicago theologian.

"Scientists are communicating directly with the public," according to a New York literary agent and author. They are "rendering visible the deeper meanings of our lives, redefining who and what we are in terms of our species, the planet, the biosphere, and the cosmos."

So let's rejoice in "the deeper meanings of our lives" revealed by science. Sure, we're just talking animals, biological machines, basically like chimpanzees and dogs, but much, much smarter. In the cosmic picture, we're specks and dots flashing on and off like pretty little fireflies in the glorious, eternal spacetime continuum.

But why aren't science's revelations of "who and what we are" making us feel good in the real world? Why all this gloom about the future of life, this narcissism flooding the human mind and spirit, this blank indifference to anything but self-centered needs and concerns?

Come on, Science, perk us up. Surely, there must be members of your community who want to serve as role models, heroes who will tell us about the importance and sacredness of life here on earth—not just what that stupid, mindless universe is doing out there in its everlasting darkness and silence.

> Give us scientific, objective, even urgent reasons why we should start defending and advancing the cause of life—not just keep on marching in locked step toward the nearest exit to oblivion.

Rebellion in Academia

In the next part of our journey in time, we'll take a closer look at the impact of false theories of reality on the cultures of the 20th Century, but we'll never be out to "get science" or diminish its vital role in the future of humanity.

Yet, as your guide on our trail to the Quo Vadis Vision, I'll insist again and again that we, the people, are smart enough to evaluate any scientific story and then assert our right and responsibility to judge for ourselves whether the story tells us anything worthy of our attention.

When we passively accept anything we're told, science and technology change their roles from servants to masters of the human race, wipe out the beliefs and myths that give a meaning and purpose to life and, sooner or later, lead mankind on the final trail to self-destruction.

Philosophers and other scholars can also reclaim their rights to independent views, now that the Time Key has opened up fatal flaws in relativistic and quantum theories of what exists outside the human mind. Sociologist Andrew Pickering takes this view of science's "claim to truth and objectivity" in his book *Constructing Quarks*:

> There is no obligation upon anyone framing a view of the world to take account of what 20th-century science has to say.

One group of scholars at the University of Edinburgh contends that "scientific knowledge is only a communal belief system with a dubious grasp on reality."

Unifying Knowledge

The deepest human understanding of the origin and nature of our existence will eventually come from the combined studies of science and the humanities, and to achieve that goal, Harvard's Edward O. Wilson is proposing a "postmodern unification of all forms of secular knowledge." It would be a rebirth of the coalition founded during the Enlightenment period in the 17th and 18th centuries, when people explored mysteries of existence "with confidence, optimism, eyes to the horizon."

Such a coalition is more urgent in our era as physicists plunge deeper into the unknown and unknowable with little or no awareness of how psychological and cultural factors shape their theories of what they think exists outside the human mind. Their highly specialized training and experience don't cover a simple, basic fact that I mentioned earlier in connection with the meaning of space:

> When the rational mind looks out at the unknowable and incomprehensible, it can see nothing more than reflections of itself. Only the "eye of faith" can catch glimpses of anything that may exist beyond the boundaries of the intellect.

So what do we see now in our lives and the universe after a century of revelations from "scientific faith"?

Discussing Reality

The *knowable* secrets of the universe are hidden in the cognitive coupling between mind and reality, and a broad range of disciplines will be needed to differentiate conceptual possibilities from what actually exists in the physical world outside the mind.

(In a truly "enlightened age," cosmologists would have conferred with psychiatrists or therapists before presenting the world with a theory like the Big Bang.)

A productive search for Wilson's "fundamental unity underlying all knowledge" will begin when philosophy, psychology, sociology, history and other disciplines reclaim their positions as equal participants with science in all secular discourses on the origin and nature of our existence.

Most people want more advanced technologies to enhance their lives and give them a firmer grasp on the world around them. But after a century of trying to make sense out of the reality created by what Einstein himself finally recognized as the "scientific and technical mentality," and struggling to hang onto a jumble of shattered values and beliefs, don't people today have a much more urgent need for a renewed faith in life and all its great possibilities?

The gate to those possibilities will open when the academic community unites with the people in declaring they have had enough of the reality imposed on them by Einstein and the quantum theorists who branched into the philosophical, cultural and spiritual realms of the human condition.

20

WHO'S DREAMING HERE?

> Until recently, the scientific community was so powerful that it could afford to ignore its critics—but no longer.
>
> *Scientific American* editorial

In looking at trends likely to extend into the new millennium, editors of *Scientific American* also noted that a growing number of scholars now challenge whether science has a legitimate claim to truth and objectivity.

Even ordinary people, the magazine said, have joined in disputing scientific "truths." However, those people make "ridiculous assertions about UFOs and the supernatural" and spend their leisure time watching movies and TV shows about sinister extraterrestrials, kindly angels, out-of-body experiences and various other "illogical and unreasonable" stories.

The editors saw "this ominous phenomenon" as a "wave of irrationalism [that] threatens to engulf society and, in the process, impede science by robbing it of support and brains suitably equipped for the rigors of future research."

To counter the threat, defenders of "Big Science" are writing articles and books and holding meetings to rally public support for more funding of that future research, much of which is focused on such phenomena as cosmic superstrings, dark matter and black holes, big bangs and big crunches, antiworlds and parallel universes of higher dimensions. In the words of physics professor Freeman Dyson:

> [Such research] does not do much harm, or much good, either to the rich or the poor. The main social benefit provided by pure science in esoteric fields is to serve as a welfare program for scientists and engineers.

Identifying Enemies

Apparently, there is a growing distrust of science, but to understand it better we should recognize that the term "science" covers a broad range of disciplines. It would be against our vital interests not to support medical science, which is working hard to enhance and extend our lives, and we should encourage and support every area of study and research that increases our knowledge of the natural world so that it becomes a trustworthy servant and friend, not an enemy that leaps out of nowhere and threatens to destroy us.

In my view, the best way to identify our "enemies" (a very small but highly influential minority in the science community) is to ask ourselves:

Who has been telling us all these years that humans and their civilizations are nothing more than momentary specks and dots in the glorious cosmos? Who has classified us as nothing more than talking animals, biological machines? Who has been elbowing their way to the pulpit for a turn at urging us to abandon our spiritual hopes and embrace the quantum view of reality?

If that is all we are, then we might as well jump into that "wave of irrationalism" and just have ourselves a jolly good time. If problems arise that we can't solve on our own, maybe we'll be "touched by an angel" who has the right answers. Or maybe one of those UFOs will swoop down and carry us off to a happier universe of higher dimensions.

Ghosts, goblins, screams in the night—let's believe in anything but a universe that doesn't know or care that we are here.

Fiction or Science?

The problem, as the theorists see it, is that we can no longer distinguish between science and science fiction. They don't object to our fascination with "pseudoscience," from psychic hotlines to *Star Wars* and other sci-fi thrillers; in fact, some physicists say they also love science fiction and they applaud the writers and producers for "educating" the public on how things work in the universes of relativistic and quantum science.

However, they are concerned that such interests distract the public from the goals and financial needs of their research projects. That's understandable but we have to ask: Who has been blurring the distinction between mind and reality?

Let's get out our existential pens and draw our own pictures of who's doing the dreaming in the world of quantum cosmology.

First, let's trace again the theory that "reversible time" is an intrinsic feature of the material universe outside our minds. The idea arose from mathematical models that could move Einstein's time in both the forward and backward directions, but let's pin it down in our Quo Vadis reality, based on logically and empirically supported concepts of space and time.

Time Travelers

When we relate the brain's motion-detection system to what's happening in the world outside our minds, we're talking about observing an ongoing sequence of events driven by four natural forces. So if we could reverse the sequence, would my cottonwood tree shrink in size and disappear into a seedling in the ground? Would a bullet pluck itself out of the victim's chest and fly back to the shooter's gun barrel? Would the dead rise from their graves and relive their lives backwards, as in a movie in reverse motion?

If a time traveler followed his life's sequence of events in the reverse direction, would he return to his mother's womb and finally disappear into bits of organic matter, thus avoiding that old paradox about him changing events that had occurred before his birth? Or if he could speed up the sequence in the forward direction, would he be merely hastening his flight into the waiting arms of old age and death?

Seems like he would be trapped in the time zone of his own life span, unless he could enter some "other dimension" that detached him from the present. But then where would he be? Floating around in empty space and *time* (i.e., his brain functions)?

Sex in Reverse

Some quantum theorists say that time "dilates" instead of traveling backward or forward in a straight line of events. Then could the time traveler, in an observer-ruled universe, control his movements in time and space by applying medications that made the pupils of his eyes expand or contract? (You might want to try that out the next time you're late for a date.)

Could people adapt to an existence in the backward mode? Wouldn't they soon starve to death if they had to eat and digest food in reverse? Might even get disgusting, don't you think?

> What would sex be like for them? If they always started with the climax, wouldn't it get rather tedious, not to mention pointless, if they had to back through all the foreplay?

Maybe this report would make more sense to people if they started with the end of the last word in the text and read backwards to the Prologue's opening sentence (*?efil htiw gnimeet esrevinu eht t'nsi yhW*).

But, for my part, I'm not sure I could retrace my brain functions and rewrite this book from finish to start.

So we have to ask:

Who introduced the world to the idea of traveling back and forth in *time*, gave it the imprimatur of a legitimate scientific theory, and now lavishes talents and resources on figuring out various designs for *time* machines?

It wasn't and isn't the science-fiction writers.

Bangs and Crunches

Now for the Big Bang theory, which perhaps more than any other raises the question of who's doing the dreaming here. The theory says the entire universe exploded out of single primordial particle, roughly a trillionth of a trillionth of a trillionth of a trillionth of the size of the *period* at the end of this sentence. Or this one. Or maybe this one. Take your pick.

In that cataclysmic creation some 15 billion years ago, the substance of the cosmos—*time*, space, matter, energy and the laws of Nature—achieved its basic form in far less time than you spent in reading this sentence—actually, a billionth of a trillionth of a trillionth of a second or so, according to some estimates.

From that quick beginning out of one miniscule particle of space and *time*, the theory says, we now have an observable universe whose total substance weighs 100 trillion trillion trillion trillion tons. Can you, if you still believe in old-fashioned rational thought and common sense, imagine packing even a spoonful of sugar into a trillionth of a trillionth of a trillionth of a trillionth of the space occupied by just one of these periods here…?

But even a trillion etcetera tons didn't seem to be enough to account for the rate of cosmic expansion, so to balance out their equations, the cosmologists concluded that up to 99% of the universe must be made of invisible "dark matter." If ordinary real matter is only 1% of the cosmos and this earth and the people around us, there'll be enough of the alien substance to provide the gravitational force needed to pull everything back into the primordial particle, and the cosmic story will end in the Big Crunch. As physics professor Thomas Weiler sees the outcome:

Dark matter is the last affront to our worth. We're not even made of the majority stuff.

> So who's doing the dreaming here? Since everything in both cosmology and science fiction has become so bizarre, the question would be difficult to answer, but now we can apply the Quo Vadis response that helps restore good old reality in our lives:

The "4th dimension of time" exists solely in the conscious mind. Space is an empty void with no properties or features that could affect the movement we observe with our mental motion sensors.

> Therefore, universes don't explode out of tiny particles of space and time.

Except in the dreams of cosmologists and science-fiction writers.

Turbulent Century

Future historians will take note of many extraordinary features that characterized the turbulent 20th Century: total wars on a world scale...mass exterminations of millions of men, women and children...mankind's acquisition of a god-like power to bring on a real Armageddon and destroy all life on this planet...the first steps in space exploration...unprecedented technological advances that revolutionized society, produced higher standards of living for more people, and began the destruction of the natural environment...the victory of democratic capitalism over fascism and communism...the emergence of terrorist movements as the gap widened between prosperous and impoverished nations...a turning away from the beliefs and myths of earlier generations and the adoption of new ideas that stripped life of any significance...the dawn of the Information Age and the Era of

Robots...the first versions of computer-generated virtual realities into which people could escape when the world around them seemed threatening, boring or meaningless.

After reviewing all of these complex, dramatic and radical changes in the human condition in the course of the 20th Century, the historians may apply one simple title to sum it all up:

> The second millennium ended with "The Century of the Big Bang."

21

Time for Conclusions

The issue is simply whether or not two plus two equals four.
Philosopher Albert Camus

As in the fable about the six little blind men who encounter a huge elephant and use their hands to try to figure out what it is, the "conceptual eyes" of the math models can be blinded by false premises on the nature of the reality outside the human mind, and so the models can only conjure up fantasies of what the universe is.

Let's write our own fable, based on the ancient Hindu belief that people and their worlds exist only in the "timeless dreams" of the gods. We, as products of dreams, aren't supposed to know that we have no reality of our own; if we did, the gods would be confronted with a paradox—who's really doing the dreaming here, us or them?

We solve the paradox by discovering the true meaning of time as a mental function of our subjective reality that gives us our sense of being alive. Now we can accost the gods and strike a bargain with them:

"You give us a separate, independent existence in a temporal world of reality, and we'll give you all of eternity to figure out how *you* can be dreaming in a *timeless* realm where absolutely *nothing* could be happening. Come up with something, you guys, or you and your worlds will be forever frozen in your dream capsules."

No More Puzzles and Paradoxes

Nature did not and never will have a purpose for time except as a motion sensor in the mind of life. We are Nature's clock, the only timekeepers in her universe. We are time. Time is us. What idea could be more "utterly simple"—and simplifying? As physicist John Wheeler might say after reading this report on our journey in time:

Oh, how beautiful. How could it have been otherwise?

We have solved science's "welter of puzzles and paradoxes" by recognizing and defining two separate but closely related realities:

> There's the mental world inside our minds, where the brain's motion-detection system keeps track of what's happening in our lives, and then there's the material world outside our minds, where everything is controlled by four natural forces.

There may be other natural forces yet to be discovered, but I'll bet any physicist—how about the next round of drinks at his or her favorite club?—that time won't be among them.

As for "science's greatest mystery," the "arrow of time" also cannot exist, except in our minds. If you followed a sequence of taking three steps forward and then backing up three steps, some "arrow in space and time" wouldn't flip back and forth to indicate or control your movements—even if your feet moved at near the speed of light, relative to an observer (I'll explain Einstein's speed qualification later).

We use the "arrows" in our minds to follow the course of change, which—from the atomic to the celestial level—Nature manages very nicely on her own without any help from our equations and the brain clock she gave us so that we can witness what her skeleton is doing.

And when we humbly and carefully observe the real world, we can come up with principles, such as the Second Law of Thermodynamics, that help us understand how and why Nature operates her universe in "one direction of time."

What should our Quo Vadis reality mean to theoretical physics?

> It should mean that the focus of theories and research can turn back to a physical reality in which causality—not exotic equations, brain functions and time metaphors—determines and controls what's happening in the non-conscious material world outside our minds.

Ready for This?

Is that existential line still clearly drawn between the mind and reality? Okay, let's take a few more swings at conventional theories of what's happening outside our minds.

What we call time is a sophisticated, biological motion-detection system in the brain. It is not a physical substance or dimension embedded in space.

Therefore, a universe now filled with 200 billion galaxies (the official number when this report was written) didn't slip through a rupture in the fabric of space and time. How could a "fabric" or anything else be made out of brain functions woven into the spatial void?

(Do you suppose that space, like time, is some kind of intelligent entity that responds to our thoughts and equations? Or maybe the tuxedo is the universe after all!)

Neither did a hundred billion trillion baby stars explode out of an infinitesimal particle, as in the Big Bang theory, and instantly create their own space and *time* (i.e., stellar brain functions).

Exciting News for Students

Hundreds of thousands of science students and their teachers should have some exciting classroom discussions when the news gets out that Einstein's spacetime continuum, the core of 20th Century theoretical physics, is actually somebody's brain functions clinging to empty space.

What will happen to the universe if we take away its own motion sensors? Will it freeze up? Vanish in a puff of stellar detritus? Or will it just keep on rolling along?

Good News for Authors and Publishers

They'll be busy replacing millions of outdated school textbooks, CD-ROMS and popular science books. Also many volumes on the philosophies, history and cultures of the 20th Century.

Who can believe anymore in theories and other worlds rising out of this weird mixture of space and *time*?

Bad News for Sci-Fi Writers

Relativistic effects predicted by Einstein's equations—such as time speeding up or slowing down as space shrinks or expands—don't kick in until the velocity of an object like a spaceship approaches the speed of light (relative to an observer).

But how could the motion of a spaceship, slow or fast, affect time, which exists only as a timekeeping system in the observer's mind? Wouldn't he get dizzy, or even go mad, if his brain clock had to tick faster or slower (or even run backwards) whenever a spaceship, baseball, molecule or other object whizzed past him?

Looks like we'll have to junk the warp drives and time tunnels that starships have been using to travel around the universe. Gosh, I hope science-fiction writers can come up with another bold way to transport their characters to places where no one has gone before.

Not to worry, Capt. Jean-Luc Picard. You can keep your appointment with the Klingon ambassador by traveling to his distant galaxy during those long commercial breaks.

Worse News for Particle Physics

Researchers at the U.S. National Institute of Science and Technology will have to reassess their recent report that "atoms can exist in two different places *at the same time*." [emphasis added]

On the other hand, results of the government's project confirm my suspicions that a certain woman, who is presumably made of atoms, exists in two or even more places at the same time. Well, at least in my motion-detection brain functions.

Even Worse for Cosmologists

Sorry, but antiworlds and parallel universes of higher dimensions simply cannot be constructed out of space and brain functions.

Maybe they could exist in a cosmos made of more than three *spatial* dimensions, but such universes can't be observed by humans with 3-D brains. I guess they'd have to be rigid, petrified worlds where no skeletons could be seen dancing around to the tune of the equations of the theorists.

Anybody Laughing?

Heaven forbid that I should ever be snide or haughty in telling you about the ideas of the revered authorities. In the minds of some people, speaking ill of science is more offensive than hurling curses at Almighty God. I'm just tossing out these touches of ironic humor along the trail in hopes of more effectively calling attention to the deadly serious issues and choices we're all facing now.

The laughter and snickering will end and the weeping and wailing of six-billion men, women and children will begin in those last days or hours, unless enough people step forth while there's still a chance for survival and give a bold, confident answer to the question hanging over this generation:

Quo vadis, Humans? Where are you going now?

Look into the faces of your children and you will want to shout your answer to the world.

Part Five

Lessons from History

22

"We Are Guilty"

> Forgive me, Newton; you found the only way which, in your age, was just about possible for a man of highest thought and creative power...although we now know that [your theories] will have to be replaced by others farther removed from the sphere of immediate experience.
>
> Albert Einstein

Scientists in virtually every field have earned our respect and support—anthropologists, chemists, biologists, geologists, astronomers, medical researchers, entomologists, mainstream physicists, archaeologists—all of the men and women who are working to cure our human ills, extend our life spans, bring the fruits of this earth to the poor and homeless, develop faster and safer modes of transportation, increase our knowledge of the natural world in which we live.

We should also extend our praise and support to the space pioneers who want to expand the human presence into our solar system and beyond—an enterprise that could someday bring us into closer contact with the larger universe and inspire a bolder, grander vision of the future of humanity.

The late Carl Sagan, one of my favorite stars in the science firmament, was an eloquent advocate of that vision. He argued that the advent of space exploration brings us to a fork in the road. If we choose to confine our lives and work to this home planet, then our fate will be decided, either sooner or later, as the fragile biosphere of life degrades and we edge closer to a nuclear holocaust as nations compete for dwindling space and natural resources.

The other branch of the road could lead us into a future in which all peoples on this earth would be united in the highest mission that humans can conceive for themselves.

Regrettably now, at least for those who share Sagan's vision, we didn't aggressively follow in the footsteps of astronauts Neil Armstrong and Edwin ("Buzz") Aldrin who first landed on the moon and took "one small step for man, one giant leap for mankind." If we had pursued a much more ambitious program since then, we would likely have several large space stations orbiting the earth and perhaps other planets by now, a permanent outpost on the moon, and we might even have established a small colony on Mars. As Sagan said in *Cosmos*, his classic PBS television series:

> We have lingered long enough on the shores of the cosmic ocean. We are ready at last to set sail for the stars.

Waiting for Messages

The Big Bang theory, which estimated the universe's age at about 15 billion years, was based on impossible premises, so now astronomers can turn to other factors as probable indicators, such as the birthrate and life span of stars and galaxies. A reasonable guess may be that the active universe has been around for trillions of years; in probability terms, that would make the odds extremely small that we and all other life forms would appear only near the beginning or the end of cosmic time. We would be much more likely to land somewhere in the middle section of the scale, which would give us plenty of time to develop the technologies needed to reach out to the stars before the cosmic curtain closes—if it ever does.

The story of mankind began about 250,000 years ago with the appearance of the first Homo sapiens; our immediate ancestors started organizing civilizations in southern Mesopotamia in about 3,500 B.C.; most of our technologies emerged in the past two centuries, and the rate of development has been rising exponentially. So once intelligent life establishes a foothold in a favorable environment, it doesn't take long to build an advanced civilization.

> In just another century, we could colonize parts of our solar system and start sending out manned and robotic spacecraft to explore more distant frontiers. Think of what our descendants could accomplish if they had a thousand more years to fulfill a higher mission for life. If our species survives and flourishes for another million years, surely it will populate a large sector of the universe.

So why aren't we receiving any messages from intelligent life forms that must have appeared millions, billions or possibly trillions of years before us?

Cosmic Chorus

Personally, I couldn't believe in UFOs unless alien beings appeared before the whole world or broke the cosmic silence with a clear message. However, it seems highly probable that other forms of intelligent life do exist somewhere in this vast universe, and astronomers have begun to find evidence of solar systems in the Milky Way and other galaxies that have trillions of stars just like our sun.

From the optimistic view, it's intriguing to imagine that we are still babies in the cosmic story and that more mature beings visited our planet in past ages. Perhaps they intervened on occasion in human affairs but then decided we hadn't advanced enough yet to justify continued contact.

It's even more intriguing—and inspiring—to hope that our descendants will be the wise and benevolent sojourners who try to help other life forms add their voices to a cosmic chorus.

Which brings us back to the big question: How do we, at this stage of history, develop the values and beliefs needed to survive and flourish, now that we have the power to self-destruct?

Giving up that power is not an option. Except for a few who retreat into their own private Shangri-Las, intelligent beings will keep climbing the ladder of science and technology until they reach whatever destiny awaits them.

> Humans want to know who and what they are and why they exist, and life-affirming answers can come only from an expanding knowledge of the physical reality from which living beings arise—combined with principles, goals and passions that could give life a chance to end the cosmic silence forever.

Or will we, like all other forms of intelligent life before us and possibly out there now, rise only to a limited height and then fall back into oblivion?

The perennial—and hardest—first step has always been to learn from mistakes of the past so that we aren't doomed to keep repeating them. Let's briefly review the history of ideas that shaped the worldviews of the 20th Century and see what we can learn, starting with the work of Einstein, the preeminent scientist who introduced us to a radically new and different vision of the physical reality outside our minds.

Spiral of Pure Thoughts

I never met Einstein but he was a familiar figure in my boyhood fancies. He was the revered physicist who had reconceived the universe, and for me and many others, he was the role model who stirred ambitions to one day join the next generation of scientists inspired by his visions. We would take up his famous equation $E=mc^2$, raise it as a mighty sword and fight our way ever deeper into the mysteries of the universe.

He became my good friend, in those daydreams, although I always addressed him as "Doctor Einstein." We had many discussions on practical as well as scientific matters, and I often turned to him for advice when I felt overwhelmed by the problems of growing up. I even sought his insights into the peculiar behavior and feelings of girls. *You must be patient, Robert. Becky needs more time and space to deduce the relativity of merit among the boys clamoring for her attention.*

In seven years as a student of theoretical physics and math, my boyhood image of Einstein as a mentor gradually faded as I began to feel a stronger need for a reality I could believe in and understand. Einstein's approach to finding "the truth" led in a direction that I no longer wanted to follow.

However, I have never lost my love and respect for Dr. Einstein, the noble man and great thinker. There is a quality of innocence in his theories, a purity in his thoughts and feelings. He was a proud, at times arrogant, man but he had a rich sense of humor and he laughed and joked about the fame and honors descending on him. He placed little value on material possessions except for his pipe and the violin he often played to soothe his spirit and inspire his thoughts.

> In his early work, Einstein believed that "pure thoughts can grasp reality as the ancients dreamed," but later concluded that "pure logical thinking cannot yield us any knowledge of the empirical world. *All knowledge of reality starts from experience and ends with it.*" [emphasis added]

By then, however, it was too late to halt the spiral of "pure thoughts" into antiworlds and parallel universes of higher dimensions of space and time. Before his death in 1955, Einstein, the "philosopher of a new, enlightened age," accepted his own share of the blame for what he called the "dehumanization and mechanization" of human beings and their societies.

"We are guilty," he said. "Man grows cold faster than the planet he inhabits."

23

Age of Relativism

> I want to know how God created this world. I am not interested in this or that phenomenon. I want to know his thoughts, the rest are details.
>
> Albert Einstein

Einstein was said to work more like an artist than a scientist, and since, contrary to the claims of some biographers, he was a deeply spiritual man, he sought his ideas in "the mind of God."

He rejected the authority of his Newtonian predecessors, who were guided by the empirical method, and relied instead on the power of his intuition and deductive reasoning. He once said that imagination (the subjective) is more important than knowledge (the objective). Testing his theories through experiments and observation was merely "detail work" for others.

"Einstein felt he had a direct pipeline to God," according to one of his followers, and he had a mission to discover a universe that matched his lofty vision of the beauty, simplicity and symmetry of Nature.

In what one writer called "the most profound event in the history of human thought," Einstein conceived a four-dimensional universe structured and operating in a spacetime continuum.

Modes of Thinking

Once freed of the constraints of Newtonian science, other physics theorists employed their own intuitions in expanding Einstein's view of space and time into new regions of quantum mechanics and cosmology. Soon the longstanding connection between cause and effect was broken, and random chance and uncertainty took over as the ruling principles. Answers to why and how things happen became "indeterminable," and therefore, the theorists said, the questions were "meaningless."

> By discarding the empirical methods of the past, Einstein ushered in the Age of Intuitive Physics, and in the course of the 20th Century, the distinction between mind and reality gradually faded into what one scientist called the "merest shadows."

Equations were no longer useful tools in defining the observable universe; they had the power to *create* new universes. Symbolic models no longer *represented* reality; they *were* reality. The map was the territory, the menu was the meal.

In such realities, theorists encountered what physicist Thomas Brody described as "a disconcerting variety of different and incompatible forms, such as multivalued logics, modal logic, the logic of relevance, quantum logics and intuitionist logic." As a result, Brody concluded, "the simple identification of what is rational with what is logical...must be abandoned; but we do not know yet what should replace it."

The theorists still do not know what should replace the rational, conscious-reasoning approach in scientific thought.

Nothing Is Certain

The new physics began to influence the cultural world in the 1920s, when quantum pioneer Werner Heisenberg impressed a new message on human consciousness: *Nothing is certain*. Not on the "microscopic" level of atoms and molecules. Not in the "macroscopic" world we see around us and in the heavens. Not in our personal and social beliefs. Not in our morals and motives.

Everything is relative.

Moral relativism led eventually to an unraveling of the social order, a pervasive cynicism, a distrust of all political and cultural authority, and what one commentator called "a chaos of manners and speech, and new varieties of decadence." Poets and philosophers, such as T.S. Eliot and André Malraux, retreated to the "religion of art in order not to die of the truth" revealed by science.

Polls indicate that the majority of people still believe in the existence of God, but self-centered needs and values, rather than the rules of traditional faith, more often determine motives and choices in the business, cultural and political worlds.

The emerging trend alarmed some of Einstein's colleagues and drove a few of them into despair. The Dutch physicist H.A. Lorentz, from whom Einstein borrowed the basic equations used in his early theories, told a friend in 1924:

> I have lost the certitude that my scientific work was bringing me closer to objective truth, and I no longer know why I continue to live. I am only sorry not to have died five years ago, when everything seemed clear.

Back to the Dark Ages

Eventually, Einstein expressed dismay over the radical new directions in physics. "God does not play dice," he admonished his followers. But no one wanted to turn back to Isaac Newton's rational, well-ordered, comprehensible universe. "Go fishing, old man," one physicist said in dismissing the master's protests.

More recently, skeptical physicists have expressed uneasiness over science's intrusion into the spiritual realms of our cultures. Nobel laureate Sheldon Glashow noted in 1987:

> [The contemplations of quantum theorists] may evolve into an activity to be conducted at schools of divinity by future equivalents of medieval theologians. For the first time since the Dark Ages, we can see how our noble search may end with faith replacing science again.

However, some scientists still assert that they must take on the dominant role in defining our moral values and spiritual hopes. They believe that science, not religion, offers the surest path to God, who is variously identified as a quantum cosmologist, a divine supercomputer, or the Omega Point.

In one view openly scorned by many scientists, Frank J. Tipler, a professor of mathematical physics at Tulane University, says the scientific deity will emerge from the evolutionary process in the coming aeons. Eventually, the deity will acquire enough power (from black holes and the Big Crunch) to raise the dead from all ages past and bestow many blessings on them, such as matching up each resurrected male (presumably including Hitler, Stalin and Jack the Ripper) "with the most beautiful woman whose existence is logically possible."

Tipler says his vision arose directly from Einstein's theories of relativity, and he offers mathematical "proof" of the salvation that awaits us all, if we can just hold out for billions of years in our graves.

Fleeing From Life

In the Age of Reason celebrated by Voltaire in the 18th Century, men believed they would eventually discover all there is to know about the universe. As the knowledge was acquired, they thought it would give mankind the power to realize all of its best hopes and dreams. They believed that human dignity and worth had an independent, objective existence as part of a rational and benevolent universe.

Now, at the turn of the 21st Century, some people still walk the old paths of hope, but for many others the human landscape has become hostile and bleak, a place that writer David Daiches characterized as "without belief, without value, without meaning," a world of "shape without form, gesture without motion." In short, T.S. Eliot's *Waste Land.*

It's like the aftermath of a storm; everything solid and substantial and believable has been blown away, and we see these sorcerers emerging from the rubble. Come, they say, crawl through our time tunnels and wormholes and you will find refuge in our antiworlds and parallel universes of higher dimensions of space and time.

In a book published in 1954, the year before his death, Einstein expressed his own despair over humanity's attempts to reconcile the clash between the old and new realities:

> One of the strongest motives that lead men to art and science is flight from everyday life with its painful harshness and wretched dreariness.... Man seeks to form for himself a simplified and lucid image of the world and so to overcome the world of experience by striving to replace it to some extent with that image.

24

The Fate of Failures

> We have seen in every country a dissolution, a weakening of the bonds, a challenge to those principles, a decay of faith, an abridgement of hope, on which the structure and ultimate existence of civilized society depends.
>
> Winston Churchill

The 14th Century stands out in history as the period in which wars, famines and plagues wiped out two-thirds of the world's population. The survivors went on to build more advanced civilizations but failed to develop a system of common values and beliefs that could restrain the inevitable dehumanization from rapidly growing technologies and industrialization. Instead, fixation on tribal, national and commercial goals continued to fan the ever more powerful flames of war and conquest.

In the 20th Century, World War I confronted mankind with the first possibility that the cycle of destruction and rebuilding was coming to an end, that all of the *quo vadis* questions would soon be answered. Nations poured their industrial and human resources onto the battlefields of

Europe, and millions of human beings fought like savage insects for possession of long trenches dug in the earth.

As the image of man as a noble creature faded, people began to lose faith in governments, religions and social institutions. Intellectuals turned away from Hegel's conclusion that "all of reality can be encompassed in a rational structure," and artists portrayed the bleak essence of the human condition in abstract symbolism, such as Dadaism and Cubism.

On the spiritual side, German theologian Dietrich Bonhoeffer raised again the Nietzschean banner, which proclaims that "God is dead, we killed Him" and therefore He can no longer be the source of meaning and hope. The philosophy of existentialist despair seeped into the theories of German and other scientists.

> Human existence is an absurdity, the prophets of meaninglessness declared. Yes, the pioneers of relativity and quantum physics echoed, and we can prove it scientifically.

The best image that influential writers could offer was "man standing up bravely before a meaningless sky."

River of Nihilism

German historian Oswald Spengler, in his 1918 book, *The Decline of the West,* saw another cycle of "annihilating doubt" spreading through science and society. He singled out Einstein's substitution of abstract models for the real universe of "hopeful earlier generations" and denounced it as a symbol of the dehumanizing forces that would bring about the disintegration of Western civilization.

The Great Depression in the 1930s, followed by another total war of even greater horrors, completed the merging of cultural and scientific thought into one river of nihilism. For many young idealists in the postwar years, the new quantum philosophy confirmed what they were prepared to believe about politicians, the police and legal systems—a distrust of all authority that verges on paranoia in the minds of many people today.

The dominant impulse since the Beat Generation of the 1950s and later youth cultures has been to deconstruct the ideals and institutions that had, however imperfectly, guided individual and group behavior in the past. For many children growing up in the Sixties and later, this "Yippie" declaration in Chicago during the 1968 Democratic National Convention expressed what they felt was the only course to a better world:

> Break down the family, church, nation, city, economy; turn life into an art form, a theater of the soul and a theater of the future. What's needed is a generation of people who are freaky, crazy, irrational, sexy, angry, irreligious, childish and mad.

Now attempts by "extremists" to bring back the old ideals, or establish any other popular consensus on moral and ethical values, are rejected as biased against a particular group or system of beliefs or nonbeliefs.

The loss of faith has left society with two common anxieties, according to Harvard professor Michael J. Sandel:

> One is the fear that, individually and collectively, we are losing control of the forces that govern our lives. The other is the sense that, from family to neighborhood to nation, the moral fabric of community is unraveling around us.

Sewer Culture

Ideas create the realities in which we live; images in our minds make us into the beasts of war, or inspire us to work and sacrifice for the good of others. But, like preceding generations, we could not find a guiding light to lead us through the wilderness of a strange new world. We transformed the rigid conventions of the Victorian era into the opposite extreme of anything-goes, nothing-but-me matters—and now we live and die in a sewer culture where human conversation flows in currents of casual obscenities, where the cries of the victims drown out the calls for personal responsibility, where drugs and violence soothe the guilt and fears of failure, and where man, once the measure of all things, asserts his standing in the universe by how well he performs the mating rituals of dogs and barnyard animals.

Life is hard, then you die. Nothing you do matters. Here today, gone tomorrow. Yours is not to wonder why; yours is but to do *and* die.

What happened to human pride and dignity? Willard Gaylin, a New York psychiatrist and a lonely advocate of the human cause, puts it this way:

> We don't much like ourselves these days. We are a failure in our own eyes, and that is always a dangerous and unstable state, whether for the individual or the species.

As failures we must discard this "pompous, egocentric nonsense" that we still hold a superior position in the hierarchy of creatures on this planet, says one animal-rights activist. Human life has no intrinsic value, and if the mental or physical state of a man or woman or child falls below a certain "quality"—a standard that will no doubt be determined by a "committee of experts"—why then such useless creatures must be quietly dispatched.

They shoot horses, don't they?

25

END THE CRAZINESS

I am mad as hell, and I'm not going to take this anymore!
Actor Peter Finch in *Network*

We have been using our reason and common sense to decide for ourselves what's believable in the stories of science. Cosmologists and quantum theorists themselves describe their versions as "bizarre...hard to believe...the craziest idea that anyone could ever come up with." Even quantum pioneer Niels Bohr reacted with this exclamation when told of the latest theory in his field of science:

"It's not crazy enough—it can't be right!"

To be doubly sure we aren't telling crazy stories too, let's go over once more how the blurring between mind and reality can occur in human thought and scramble the symbiotic relationship between those two separate worlds. Since our own thoughts may still falter when we differentiate between time and physical reality, or try to recognize and accept the fact that the world before our eyes is shrouded in perpetual and eternal darkness, perhaps another quick review will help us maintain a clear vision of the fortress of hope and meaning we can build in our minds.

When we observe something, or think or talk about it, we apply "word-concepts" that connect immediate perceptions with past data recorded in our memories. Look at that four-legged object with the flat surface in your dining room. It's a "table." The word-concept exists only in your mind; the table exists only in the physical world outside your mind, and you have no difficulty in distinguishing between the subjective and objective.

The word is not the table. The table is not the word.

"Ol' Man River," always one of my favorite songs, blends concepts of "man" and "river," and Jerome Kern's flowing music and the singer's deep ancestral voice use the combination to evoke powerful images that help us reflect on human mortality; otherwise, as the distinguished mythologist Joseph Campbell noted, "the river is just water rolling along."

The river is not an old man. An old man is not the river.

We apply word-concepts to everything we experience, and in some areas the distinction between words and reality do become blurred. Love, for example, and hate are word-concepts that represent a complex weaving of perceptions, emotions and instincts, and often we can't tell the difference between what exists in our minds and in the real world.

But now, in our quest for a viable, believable meaning and purpose in life, we can use our existential pen to draw a clear line between mind and reality: Time is a word-concept we use in thinking and talking about motion and the changes it produces in our lives.

Time=subjective reality. Motion=objective reality.

The time is not the motion. The motion is not the time.

When we have clearly distinguished between mind and reality, we can rid ourselves of all this "craziness" and recognize that we, as humans, are the conscious, thinking beings who create and control time and all of the other senses of the mind.

We are time. Time is us.

> Let's reject, finally and totally, the use of time to construct nihilistic worlds in which there can be no fortresses "against the dark's insidious despair."

Reclaiming Our Property

In drafting his theories, Einstein innocently opened the door to a century in which the word became the table, the old man became the river, and "the very mind of reason" was shattered by "absurdities."

First, he adopted the ancient notion of time as a malleable material, a fluid substance flowing like a mighty river through the universe, and then he deduced that Nature or God must have mixed the substance together with space to produce his "uniform four-dimensional continuum."

Second, he thought of time as a kind of physical dimensional "stick" that had bonded to the dimensional sticks of space, an idea that gave his equations the power to form "warps and ripples" in the continuum's "fabric"—paths that planets and stars, along with gravity and light, would thereafter follow in travels around the universe.

But time is not a "thing like St. Paul's Cathedral"; it is not a dimension that imposes geometrical structures on the physical reality outside our minds.

> Time, in all its forms, is a property—and the *exclusive* property—of the conscious mind.

Einstein's concept of physical reality has no counterpart in any real universe. None at all. From baseballs to galaxies and the universe as a whole, motion and change are determined and controlled by an intricate network of cause and effect—not by equations and our brain functions.

Nature gave us our timekeeping system so we could function as human beings and, hopefully, continue to survive in the material world outside our minds.

As we all know, that skeleton can turn hostile and deadly when we hide his realities under self-serving illusions.

Choosing Between Trails

In various ways, we have answered Augustine's question: "What then, is time?"

Time is the neurobiological system in the brain that processes thoughts, feelings and sensory information, and causes and drives a sense of continuous, ongoing consciousness in the mind.

Our Quo Vadis trail leads to a reality that exalts and ennobles life, and it is marked at every step by logic, good old common sense and a growing body of scientific evidence.

For their part, theorists should reevaluate the evidence they found on their own trail to reality: Time is a substance that flows to infinity, a dimension that bonds to space, and a clock-like force that regulates motion in the material world.

We did our own review of their evidence, and perhaps physicists who now see problems in a time-driven universe will dismantle the old metaphysical paradigms, preserve any pieces that still make sense to the rational mind, and start building a new model.

Whatever the reaction in the scientific community, we have our own rational grounds for making this judgment:

> Einstein's theories of relativity are idealized models that can, in limited ways, be applied mathematically in certain areas of astronomy and in research at the atomic and molecular levels, but they have no metaphysical or spiritual relevance to who and what we are and why we exist.

Albert Einstein was indeed the philosopher who "changed not only our concept of the universe, but our everyday life."

He was wrong. Tragically wrong.

Insidious Powers

Any of the observed features of the universe can be expressed in equations. That's great. There's nothing wrong or evil in mathematics itself. Science and technology, which are transforming our lives and could someday give human pioneers the power to reach out to the stars, would be crippled without mathematics.

> Time is solely a complex motion-detection system in the mind of life. Therefore, Einstein's space-time reality exists only in the minds and equations of the theorists.

Objective, realistic science and technology can be the people's great allies in the quest to fulfill the highest and noblest aspirations of humanity.

> Dimensions can define space and time, but dimensions have no separate existence outside the human mind. Therefore, quantum worlds of "higher dimensions" do not exist outside the minds of the theorists.

But if the ideas generated by the equations don't make sense to the rational mind; if they don't give us a clear, believable picture of a reality in which we can work together to preserve life, then the equations should be recognized as nothing more than mathematical tools for possible use in metaphysical-free technologies.

> False theories of reality have eroded human self-esteem and devastated traditional values and beliefs, giving us nothing spiritual in return.

We must stop giving equations this insidious power to fabricate our realities. When they tell us that our universe exploded out of a pinhead of space and time, or that humans and their civilizations are nothing more than insignificant specks and dots in the glorious quantum cosmos, we must say, no way, Jose, not a chance, get lost.

> Mathematics has no numbers or symbols, no laws or principles, that can satisfy the human longing for a higher meaning and purpose in life.

A fresh vision of humans beings as the supreme achievement of the universe will inspire people to do whatever it takes to preserve life on this planet and work together to someday achieve a higher destiny as explorers of new worlds in the cosmos.

Part Six

The Quo Vadis Vision

26

Searching for Hope

> The great mystery is not that we should have been thrown down here at random between the profusion of matter and that of the stars; it is that, from our very prison, we should draw from our own selves images powerful enough to deny our nothingness.
>
> French philosopher André Malraux

According to ancient legends, the gods, back at the beginning of human existence, decreed that fire and light would henceforth be confined to their majestic home in the heavens. Man, abandoned on earth, must wander around in darkness unable to acquire any knowledge (science, or *sciens* in Latin) of his world.

Then Prometheus, a god who sympathized with the plight of the human creatures, stole the heavenly fire and cast it down on earth, giving man the power to see and understand. In retribution, the supreme god Zeus chained Prometheus to a rock and condemned him to an eternity of torment by vultures that ate his liver every night.

Inspired by such legends, man ruled the universe through his gods for thousands of years. If at times he lost control—the weather got nasty, grasshoppers ate up his crops, an earthquake wrecked his cities, he lost a crucial battle with an enemy—he thought he must have offended the gods. However, he could placate the angry deities through elaborate rituals and sacrifices, and thus regain control of his destiny.

Then natural philosophers began to notice that the universe operated pretty much as it wished and didn't really need any help from man and his gods. Man's role in the cosmic scheme of things was reduced to a thoughtful witness who observed events and tried to figure out the natural laws that kept things going in an orderly, predictable fashion.

In the 17th Century, Isaac Newton gathered up what he and his predecessors had observed and, in his 1687 *Principia Mathematica*, he presented humanity with a comprehensive view of how things work in the real world. His natural laws remain valid and trustworthy to this day in every area of empirical science and engineering, from building bridges to space exploration.

On the side of subjective reality, Bertrand Russell and other philosophers interpreted Newton's universe as a melancholy place where determinism had ruled out free will and the traditional aspirations of mankind. Others welcomed a rational, comprehensible world in which humans could function more effectively in building advanced civilizations that would bring the blessings of life to more people.

> They could know with certainty that *something* was real, that their lives were firmly attached to a solid framework of reality, and when they felt a desire for something more, they could venture out to the shores of the unknown and unknowable and look for signs of a higher meaning.

A Fork in the Road

Inevitably, in a complex and mysterious universe, puzzles and contradictions arose that couldn't be explained adequately by empirical science. In the first part of the 20th Century, science and humanity reached another of those historic forks in the road.

Newton and the other old savants beckoned from the road leading off to the right. Come, they said, follow our path and it will lead you deeper into a rational, comprehensible universe. The puzzles confronting you now can be solved through more careful observations and ingenious experiments. Go forward in the pursuit of objective knowledge.

From that road off to the left, we heard the pioneers of relativity and quantum theories calling to us. Come this way, they said, and we will lead you into a new and more fascinating land that lies far beyond the boundaries of your poor human senses and reason.

Newton's trail could well have led to a more rapidly expanding knowledge of the mechanisms and strategies that Nature evolved on her own without man's help, and from that knowledge, technologies beyond anything we have today.

> Nonpolluting, safe, virtually limitless sources of energy, the primary force behind advanced civilizations. An end to smog and other toxins that are poisoning life and the environment. The power and resources to transform blighted cities, to make poverty and disease rare phenomena, and to begin to reach out to the stars.

Instead, we took the intuitive road of the theorists and wound up in a place where nothing real, nothing but fantasies, will ever be built—not in antiworlds and parallel universes of higher dimensions, not in warps and ripples in the spacetime continuum. Science and humanity have come full circle, back to the belief that man has the inalienable right to design and control the universe through his gods.

Unfortunately, the designs worked out by the priests of science include no blueprints for the survival of the human species on this planet.

Reaching for Certainty

Does it really matter whether we and our universe exist in Descartes' luminiferous aether, Newton's spatial void, Einstein's spacetime continuum, or the quantum vacuum? Whether things are made of 92 natural elements or composed mostly of alien dark matter? Whether our reality is what we see and experience, or what the arcane equations of the theorists tell us?

Yes, I believe it does, because such ideas have shaped a world increasingly dominated by bizarre, nihilistic realities. When the rational and comprehensible are thrown out, the preposterous flows in and fills the human mind and spirit.

Every period of history is a struggle to adapt our beliefs and instincts to changing knowledge of the world around us, and to work it all into our cultures. The more successful periods, in terms of less cruelty and violence, more civility and compassion, were guided by a set of certainties—a social consensus on the importance and sacredness of life, on moral and ethical principles, on what is acceptable in the public behavior of individuals and groups. Humans long for certainty, whether it relates to their next meal or to what happens to them after they die.

> We can follow the road of atheism and reach the *certain* belief that death will be the final end for us. We can walk the paths of faith that lead to a *certain* conviction that life on this earth is a prelude to a greater experience. We can help maintain a stable and enduring society and die with the *certain* belief that our lives and work will be carried on by our children and grandchildren.

There is no reality in uncertainty; it is not a place where very many people can exist for long—not when science makes them doubt what their reason and senses, their best hopes and dreams, tell them is true.

The long quest for certainty can be traced in the goals that great thinkers have set for themselves over the ages. Giants of the past—Darwin,

Newton, Leonardo, Copernicus, Galileo, Kepler, back to the Greek empiricists—offered visions of a physical reality, whether right or wrong, that could be comprehended by the human mind.

And any reality people can believe in and understand with some degree of certainty, whether it arises from deterministic or spiritual beliefs or a combination of both, holds out the hope of a higher meaning and purpose in life—the first condition for the survival of the human species.

A meaningless existence is the ultimate deathtrap for all intelligent life.

27

Immortal Longings

And so we are sick for life, and cling
On earth to this nameless and shining thing.
Euripides (5th Century B.C.)

Once I asked Frank Oppenheimer, then a young physics professor and good friend, to explain Einstein's concepts of time. In the course of a long discussion in a tavern just outside the University of Minnesota campus, he basically offered three definitions:

Time is a factor in equations that predict and calculate motion; time is a substance that pervades the universe, and time is a variable force that expands or contracts space.

Then I asked Frank, the younger brother of atomic physicist J. Robert Oppenheimer, to relate those definitions to a reality I would want to believe in, either through reason or faith. After some thought, he quoted a historic remark by Herman Minkowski, a German physics professor at the Zurich Polytechnic Institute and a major contributor to the concepts and mathematics used by Einstein in drafting his 1915 General Theory of Relativity:

> From now on, space and time separately have vanished into the merest shadows, and only a sort of combination of the two preserves any reality.

The "combination" and other relativistic and quantum mixtures did not preserve any reality. Instead, they poisoned the life-affirming beliefs and myths of our ancestors, leaving us with only the despairing melancholy of the past. The dark side of the subjective overwhelmed the objective in the scientific and cultural worlds, and returned people to a reality that has always aroused instinctual fears more profound than any dread of eternal hellfire and damnation:

You are nothing and your existence is meaningless.

Therefore, you should welcome those oncoming waves of oblivion.

> To the same bourne we're driven; in the urn for all
> Death spins a lot that must erelong be cast,
> And each in Charon's boat at last
> To endless exile call.
>
> Horace (65-8 B.C.)

Faith and Science

The mind's sixth sense of time is the wellspring of the human longing for immortality. Evolution gave the early pioneers of human life the mental ability to plan ahead, imagine the possible outcomes of their efforts to find food and shelter so they could think about which future course of action would be most likely to ensure their survival and thus pass on their genes.

So far, so good. But what happened when that essential talent for advance planning reached beyond tomorrow and tomorrow and eventually bumped into the time boundaries of all mortal creatures?

Alas, that's when our ancient forebears became the first animal capable of foreseeing its own death. What a shock that must have been for them, as it is today for those of us who shrink from the idea of life ending in total and eternal oblivion.

Another survival talent had to be invoked—the ability to create myths and legends that denied human fate and gave our ancestors more pleasing images of who and what they were. They impressed their time metaphors on the universe and gave them the power to break through the barriers of death and flow like a great river into the unimaginable realms of infinity and eternity.

With the limitless power of faith, they contacted a supernatural being who dwelled in those realms—God, the Father of Time, who promised to rescue humans from the meaningless fate of mortals if they accepted his story of the origin and nature of their existence and obeyed his laws within the temporal boundaries of the life he established here on earth.

> Then Einstein plucked time from the stream of spiritual hopes and philosophical thoughts and placed it at the center of 20th Century physics. He set the stage for a science-created universe by retaining the ancient conceit that the human mind has the power to contrive and control an external form of time—and on that stage, it is man, not God, who designs and builds realities.

On the level of faith and hope, Einstein believed that his special vision would reveal a new and better reality for all of humanity in that "astonishing tomorrow." However, science and faith cannot follow the same path for long, not unless one converts the other into another version of itself.

As the old pillars of faith collapsed in the 20th Century, Einstein's followers reconceived the master's reality as a world in which the barriers of death must be rebuilt, and immortal longings quelled, and we must all face up to their vision of human reality:

We are the aliens in this universe, tiny creatures crawling over the surface of an incomprehensible being, shining our little lights of consciousness into an unfathomable dark.

> Some little talk a while of thee and me there was,
> And then more of me and thee.
> Omar Khayyám (12th Century A.D.)

Paths of Hope

Are human beings still capable of that "heroic determination to achieve a magnificent destiny"? Or have they taken the wrong turn and joined all other forms of intelligent life in a mindless march into the eternal darkness of the universe?

There are many reasons for despair, but this is a story about seeking ways to restore the importance and sacredness of life, and then to impress a life-affirming meaning on the universe. Have we found any new paths of hope that we can walk together?

Human time and consciousness are indeed the profoundest mysteries the mind can contemplate, but their meaning is not "inscrutable" as scientists say. We have, as physicist Roger Penrose predicted, started "a grand shake-up" greater than the upheavals caused by relativity and quantum theories. *We are time! Time is us!*

> ...in the dead of winter...when the sun rises at about noon, Iceland's forbidding landscape naturally reminds us of how close we all stand—the sighted and the sightless—to permanent darkness. And reminds us of the need to discover ...some fortress against the dark's insidious despair.
> Poet Brad Leithauser (1991)

Time is the seed from which conscious, thinking beings arose to become the light of the world. Time is the innate power of life that evolved human consciousness, and in a sequel to this book, I will tell you about one of many ways that theory can be applied in the search for a universe we can all understand and want to believe in.

Exploring the mysteries of mind and reality leads us to the heart of who and what we are, that inner self where humans have always sought a deeper meaning in their lives, and where they confront a basic paradox of their nature:

On one side of his nature, man seeks an omniscient, all-powerful god to whom he can bow down and entrust his destiny. "Without gods, man is nothing," says Homer's Poseidon, lord of the sea, to Odysseus, the lost wanderer. On the other side of his nature, man harbors a fierce desire to climb on his own to the highest mountaintop and seat himself on the throne.

Now we catch only rare glimpses of either of those heroic visions in the lives of our children and friends, in the world around us, and in the electronic spheres of the media and the Internet.

> The Sea of Faith
> Was once, too, at the full, and round earth's shore
> Lay like the folds of a bright girdle furled,
> But now I only hear
> Its melancholy, long, withdrawing roar....
>
> Mathew Arnold (1822-88)

If science or the gods no longer give us a meaning, we can believe that the search for meaning will become in itself the purpose of life. We can believe that human consciousness is an expression of the power of time here on earth, and when we look out at the cosmos, we can hope that humans are a sign of the universe's lonely, childlike quest to know itself and find a purpose in its own existence.

And if we believe that humans are something more than biological machines, we can trust that the immortal soul will emerge as the supreme creation of time—a state of being that is still evolving and that will remain forever hidden from science in the miracles of faith and love and hope.

Victory of Life

We have followed a trail that leads to logical, provable grounds for rejecting the metaphysical and spiritual messages of 20th Century science. If you welcome being implicated in a new conspiracy of hope, I trust you will join with other people who want to defend and advance the cause of life.

For now, grant your trail guide another turn at listing some of the meanings he sees in our Quo Vadis reality:

Our reality means we can cast off the burden of nihilism and walk along our paths without fogs of absurdities obscuring the view.

It means human beings can reclaim their position as kings of their castles and masters of their destiny, and life will no longer be the "underling" of the stars in the celestial night.

It means a place where life will be exalted and ennobled, and humans will become the master and the universe the servant.

It means humans can think of themselves again as a very special and important part of this universe, for without them and other conscious life forms, the cosmos would be nothing more than a robotic machine operating in perpetual and everlasting darkness and silence.

It means the human mind is the universe's Vessel of Time, the ship of life in which men and women will cast off the shackles of cynicism and despair and sail out on the waters of existence to fulfill the highest and noblest mission they can conceive for humanity.

It means humans are the form of life chosen by God or Nature to end the cosmic silence forever, to bring light and meaning to existence, and to someday fill the universe with their hopes and dreams.

It means men, women and children will once again speak these proud words with passion and dignity: "I am a human being!"

It means the people will turn off the "final chorus of life" and give their own bold, confident answer to the *quo vadis* question asked of all humanity in the new millennium.

> Yes, oh yes, we will survive! We will do whatever it takes to preserve the sacred gift of life and pass it on to our children and all future generations. We shall endure! We shall prevail!

Comments may be addressed to the author:

Robert J. Williams
PO Box 292075
Phelan, CA 92329-2075

E-mail address:
quovadis@cwo.com

9 780595 093304